Monographs on Endocrinology

Volume 5

Edited by

F. Gross, Heidelberg · A. Labhart, Zürich

T. Mann, Cambridge · L. T. Samuels, Salt Lake City

J. Zander, München

Jürg Müller

Regulation of Aldosterone Biosynthesis

With 19 Figures

Springer-Verlag New York · Heidelberg · Berlin 1971

Dr. Jürg Müller

Stoffwechselabteilung der Medizinischen Universitätsklinik,
Kantonsspital, Zürich/Schweiz

ISBN 0-387-05213-5 Springer-Verlag New York · Heidelberg · Berlin
ISBN 3-540-05213-5 Springer-Verlag Berlin · Heidelberg · New York

Contents

Introduction

Most of our knowledge of the physiological control of aldosterone secretion is based on animal experiments and clinical studies which were carried out in the 1950s and early 1960s by a large number of inspired, ingenious and meticulous researchers. Their work has been excellently reviewed by—among others—MULLER (1963), BLAIR-WEST et al. (1963), LARAGH and KELLY (1964), GANONG et al. (1966), MULROW (1966), DAVIS (1967) and GROSS (1967). According to the majority of these investigators, aldosterone secretion is primarily regulated by the renin-angiotensin system, with plasma sodium and potassium levels and pituitary secretion of ACTH playing important secondary roles. During the last six years, this hypothesis has been generally accepted and has only occasionally been challenged.

The following is an attempt to take—from the perspective of a relatively simple *in vitro* model—a new look at the efferent axis of the apparently very complex control system maintaining adequate aldosterone production in the mammalian organism. My views are based mainly on a series of experiments which I have performed in order to study more closely the interactions of adrenocortical tissue with substances capable of directly influencing aldosterone biosynthesis. Since all these studies were carried out *in vitro* and with rat adrenal tissue only, the information obtained by means of this particular experimental model[1] will be collated with the findings of other investigators who have used a different experimental approach to the study of aldosterone biosynthesis and its regulation in the rat as well as in other animal species and in man.

---

1 This experimental model is a modification of the *in vitro* ACTH assay of SAFFRAN and SCHALLY (1955). For detailed descriptions of methodology see MÜLLER (1965a, 1966), MÜLLER and WEICK (1967).

I. Zona Glomerulosa of the Adrenal Cortex: Source of Aldosterone

The adrenal cortex is the only organ capable of producing aldosterone in man and several mammalian species. In adrenalectomized men, sheep and dogs, as well as in patients with Addison's disease urinary excretion, secretion rate and plasma concentration of aldosterone were found to be either unmeasurably low or to be within the range of the reagent blank of the analytical method used (KLIMAN and PETERSON, 1960; BOJESEN and DEGN, 1961; KOWARSKI et al., 1964; BRODIE et al., 1967; COGHLAN and SCOGGINS, 1967a). Indirect evidence for the ectopic secretion of aldosterone by an ovarian Sertoli cell tumor in a 9-year-old girl with precocious puberty, hyperaldosteronism, hypokalemia and high blood pressure has been presented by EHRLICH et al. (1963).

Based on indirect evidence, the concept of a functional zonation of the adrenal cortex was proposed many years before the discovery of aldosterone. In a review on pituitary-adrenocortical relationship SWANN (1940) suggested that the "salt and water hormone" was secreted by the glomerular layer of the adrenal cortex, because hypophysectomy led to degeneration of the inner layers only and hardly impaired the vital "salt and water function" of the adrenal cortex. Additional evidence for the production of mineralocorticosteroids by the zona glomerulosa was found by GREEP and DEANE (1947) and DEANE et al. (1948). Potassium loading or sodium deficiency resulted in morphological and histochemical changes in the zona glomerulosa of the rat adrenal cortex which indicated increased secretory activity. Administration of deoxycorticosterone acetate or potassium deficiency caused cytological alterations which were suggestive of inactivity of the zona glomerulosa. In sodium deficiency of rats, increases in zona glomerulosa thickness were associated with an increased capacity of the adrenals to produce aldosterone *in vitro*, according to EISENSTEIN and HARTROFT (1957) and HARTROFT and EISENSTEIN (1957).

Physical separation of the zona glomerulosa from the zona fasciculata and the zona reticularis in excised rat adrenals was achieved by
"decapsulation" by GIROUD et al. (1956a). Decapsulated adrenals
produced only negligible amounts of aldosterone, whereas the "capsules" produced the same quantities of aldosterone as the complete
adrenals. Histology indicated that the capsular portion of the adrenal
gland contained the whole zona glomerulosa and approximately 20%
of the zona fasciculata. Similar results were obtained by AYRES et al.
(1956), who incubated selected slices of ox adrenal cortex consisting
either mainly of zona glomerulosa or mainly of zona fasciculata.
Per unit weight "zona glomerulosa slices" produced 500% more
aldosterone and 40% more corticosterone, but 60% less cortisol
than "zona fasciculata slices". Possibly the trace amounts of aldosterone and of cortisol produced by "zona fasciculata" and "zona
glomerulosa" tissue, respectively are due to tissue contamination or
to methodological limitations. On the available evidence it is difficult
to evaluate whether all the aldosterone is produced by the zona
glomerulosa and all the cortisol by the inner zones, or whether certain
adrenocortical cells can produce both steroid hormones. Human
"aldosterone-producing tumors", i.e. adrenocortical adenomas
surgically removed from patients with primary aldosteronism,
were generally found to contain not only high quantities of aldosterone and corticosterone, but also considerable amounts of cortisol
(NEHER, 1958; LOUIS and CONN, 1961; BIGLIERI et al., 1963; KAPLAN,
1967).

II. Pathway of Aldosterone Biosynthesis

The generally accepted sequence of biosynthetic steps involved in the production of aldosterone from cholesterol is shown in Fig. 1. Other possible pathways have been suggested by WETTSTEIN (1961), who incubated tritiated progesterone with beef adrenal homogenate and found that it was incorporated into 18-oxoprogesterone and 21-deoxyaldosterone, two possible intermediates of aldosterone biosynthesis. Tritium-labelled 18-hydroxy-11-deoxycorticosterone (18-OH-DOC) and 18-hydroxyprogesterone were converted to aldosterone by bullfrog adrenal slices (NICOLIS and ULICK, 1965). 11-Dehydrocorticosterone-4-^{14}C was converted to aldosterone by rabbit adrenal tissue (FAZEKAS and KOKAI, 1967) as well as by homogenates of monkey and frog adrenals (SHARMA, 1970). Addition of substrate amounts of 11β-hydroxyprogesterone to beef adrenal tissue slices (STACHENKO and GIROUD, 1964) and of 21-deoxyaldosterone, 11-dehydroaldosterone, or 21-hydroxypregnenolone to beef adrenal homogenate (KAHNT and NEHER, 1965) stimulated aldosterone production *in vitro*. Thus, the reactions listed in Fig. 1 can theoretically occur in a number of different sequences. However, no direct evidence has yet been presented which would indicate that activation of different pathways of aldosterone production could be of physiological importance, although this possibility has been suggested by BANIUKIEWICZ et al. (1968) and BLAIR-WEST et al. (1970b).

Labelled corticosterone is converted to aldosterone in high yield when incubated with adrenal tissue of man (MULROW and COHN, 1959), ox (AYRES et al., 1957), dog (W. W. DAVIS et al., 1968), rat (MÜLLER, 1966), bullfrog (NICOLIS and ULICK, 1965) and duck (DONALDSON et al., 1965). Based on the final specific radioactivities of corticosterone and aldosterone isolated from beef adrenal capsule strippings (zona glomerulosa) which had been incubated with trace amounts of ^{14}C-labelled corticosterone, AYRES et al. (1960) calculated that at least 50% of the aldosterone produced was derived from corticosterone. When labelled corticosterone was added in substrate

Reactions: Products:

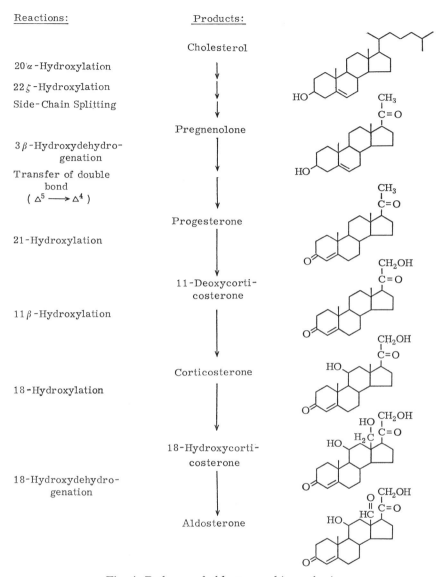

Cholesterol

20α-Hydroxylation

22ζ-Hydroxylation

Side-Chain Splitting

Pregnenolone

3β-Hydroxydehydro-
          genation

Transfer of double
      bond
    ($\Delta^5 \longrightarrow \Delta^4$)

Progesterone

21-Hydroxylation

11-Deoxycorti-
costerone

11β-Hydroxylation

Corticosterone

18-Hydroxylation

18-Hydroxycorti-
costerone

18-Hydroxydehydro-
          genation

Aldosterone

Fig. 1. Pathway of aldosterone biosynthesis

amounts, at least 92% of the aldosterone was produced via corticosterone.

On the other hand, the role of 18-hydroxylated steroids in the biosynthesis of aldosterone has not as yet been clearly established. The generally accepted assumption that 18-hydroxycorticosterone is an obligatory intermediate between corticosterone and aldosterone is mainly based on analogy and indirect evidence. Direct evidence is conflicting. Incubation of radioactive 18-hydroxycorticosterone with homogenates of bullfrog adrenal glands or bullfrog adrenal mitochondria did not result in detectable synthesis of aldosterone (PSYCHOYOS et al. 1966); under the same experimental conditions corticosterone was converted to 18-hydroxycorticosterone and aldosterone. A considerable rate of conversion of added tritiated 18-hydroxycorticosterone to aldosterone by sheep adrenal mitochondria was observed by RAMAN et al. (1966). Tritiated 18-hydroxycorticosterone was converted to aldosterone in a high yield by slices of the adrenal cortex and an adrenal tumor surgically removed from a patient with primary aldosteronism (PASQUALINI, 1964). Addition of 18-hydroxycorticosterone (but not of 18-OH-DOC or 18-hydroxyprogesterone) in substrate amounts to the medium in which rat adrenal quarters were incubated inhibited the incorporation of progesterone-^{14}C (added in trace amounts) into aldosterone (VECSEI et al., 1968). Beef, frog and human adrenal tissue slices or homogenates converted 18-hydroxycorticosterone to aldosterone (NICOLIS and ULICK, 1965; KAHNT and NEHER, 1965), but the yields of aldosterone were 15 to 80 times larger when corticosterone was incubated under the same conditions. Similarly, the rate of conversion of deoxycorticosterone and progesterone to aldosterone was markedly faster than the rate of conversion of the 18-hydroxy-derivatives of these two steroids. According to NICOLIS and ULICK (1965), the slow rate of conversion of exogenous 18-hydroxycorticosteroids could be due to the fact that in aqueous solution these compounds existed mainly as 18, 20 cyclic hemiketals and were more resistant to enzymatic dehydrogenation in this form than in the open α-ketol form. The assumption that 18-hydroxycorticosterone is the immediate precursor of aldosterone is supported by the abnormal steroid secretion pattern observed in a child with renal salt loss caused by an inborn error of aldosterone biosynthesis (ULICK et al., 1964b). The combination of severe hypoaldosteronism with hypersecretion of 18-hydroxycorticosterone and corticosterone found in this patient was most likely due to a defective dehydrogenation of 18-hydroxycorticosterone. Other cases of this rare, frequently familial syndrome have been reported by DAVID et al. (1968), RAPPAPORT et al. (1968) and TOUITOU et al. (1970).

The conversion of 18-hydroxycorticosterone to aldosterone appears to be the only biosynthetic step which is practically limited to the zona glomerulosa. 18-Hydroxycorticosterone can be produced by the zona fasciculata-reticularis as well as by the zona glomerulosa of beef and rat adrenals (SHEPPARD et al., 1963; MARUSIC and MULROW, 1967a; BANIUKIEWICZ et al., 1968). In man, 18-hydroxycorticosterone appears to be mainly a byproduct of aldosterone biosynthesis. Secretion rate of 18-hydroxycorticosterone was found to be 60 to 200% greater than secretion rate of aldosterone in normal men (ULICK et al., 1964a; ULICK and VETTER, 1965). The secretion rates of both steroids were proportionally elevated in patients with secondary aldosteronism due to cirrhosis of the liver, congestive heart failure, nephrotic syndrome or unilateral renal disease. Equal fractional increases in the production of both steroids were induced in normal subjects by sodium depletion or treatment with ACTH.

On the other hand, most of the 18-OH-DOC is derived from the zona fasciculata-reticularis—at least in the rat—and its biosynthesis appears to be mainly under ACTH control (SHEPPARD et al., 1963; CORTES et al., 1963; STACHENKO et al., 1964; BANIUKIEWICZ et al., 1968). However, small amounts of 18-OH-DOC are also produced by the zona glomerulosa (SHEPPARD et al., 1963; LUCIS et al., 1965; BANIUKIEWICZ et al., 1968). Secretion rates of 18-OH-DOC in man have not been measured yet.

Corticosterone also is produced by both the zona glomerulosa and the zona fasciculata-reticularis of beef and rat adrenal cortex (GIROUD et al., 1956a, 1958). In the rat, corticosterone is the major glucocorticosteroid (BUSH, 1953); most of the corticosterone originates from the zona fasciculata-reticularis and is produced under ACTH control (BANIUKIEWICZ et al., 1968). In man, secretion rate and plasma levels of corticosterone (PETERSON, 1956, 1959) follow closely those of cortisol, which is presumably produced only by the zona fasciculata-reticularis. According to BLEDSOE et al. (1966), approximately 90% of the corticosterone production of normal man is ACTH-dependent. When endogenous ACTH secretion was suppressed by dexamethasone, corticosterone secretion was altered proportionally to aldosterone secretion in response to changes in sodium balance. Elevated secretion rates of corticosterone were observed in some patients with primary aldosteronism but not in patients with secondary aldosteronism due to renovascular or malignant hypertension (BIGLIERI et al., 1968).

Deoxycorticosterone is the immediate precursor of corticosterone and is theoretically also of dual origin in the adrenal cortex. The capsular and the decapsulated portions of rat adrenals produce approximately the same amounts of this compound under basal conditions of incubation (VINSON and WHITEHOUSE, 1969; MÜLLER, 1971). Increased deoxycorticosterone secretion rates are characteristic of congenital deficiency of 11β-hydroxylation (EBERLEIN and BONGIOVANNI, 1955) or 17α-hydroxylation (BIGLIERI et al., 1966). In patients with these forms of congenital adrenal hyperplasia as well as in patients without primary adrenocortical disease, the production of deoxycorticosterone appears to be mainly under ACTH control. Thus, increased secretion rates of this steroid were observed in patients with increased ACTH production but not in patients with an increased activity of the renin-angiotensin system (BIGLIERI et al., 1968). The secretion rate of deoxycorticosterone was found to be normal or decreased in most patients with secondary aldosteronism associated with oedema (CRANE and HARRIS, 1970).

III. Aldosterone Biosynthesis by Cell-Free Systems

Aldosterone is synthesized by homogenized adrenocortical tissue from endogenous as well as from exogenous precursors. According to KAHNT and NEHER (1965a), the endogenous precursor of aldosterone production by beef adrenocortical tissue homogenate is most likely free cholesterol. After centrifugal fractionation of sheep (RAMAN et al., 1966), bullfrog (PSYCHOYOS et al., 1966) and rat (MARUSIC and MULROW, 1967b) adrenal homogenates, only the mitochondrial fractions were found to actively convert corticosterone to aldosterone. The enzyme systems necessary for the conversion of corticosterone to 18-hydroxycorticosterone or aldosterone could not be obtained in soluble form following ultrasonic treatment of mitochondria (RAMAN er al., 1966).

The conversion of corticosterone to aldosterone by mitochondria requires the presence of NADPH provided by the addition of NADP plus either a citric acid cycle intermediate such as fumarate or malate (TALLAN et al., 1967) or a NADPH-generating system such as glucose-6-phosphate and glucose-6-phosphate dehydrogenase. Studies on the coenzyme requirements for the conversion of exogenous 18-hydroxycorticosterone to aldosterone have yielded controversial results. According to KAHNT and NEHER (1965), exogenous 18-hydroxycorticosterone was converted to aldosterone at the same rate as exogenous corticosterone by beef adrenal homogenates incubated with NADP and NAD but without fumarate. However, according to RAMAN et. al. (1966), NADPH or a NADPH-generating system was required for the conversion of added 18-hydroxycorticosterone to aldosterone; in the presence of NADP or NAD sheep adrenal mitochondria did not convert 18-hydroxycorticosterone to aldosterone but to 18-hydroxy-11-dehydrocorticosterone.

GREENGARD et al. (1967) have presented the following evidence for the participation of cytochrome P-450 in the conversion of corticosterone to aldosterone by bullfrog adrenal mitochondria: appearance of a characteristic absorption spectrum upon reduction and addition of carbon monoxide, inhibition of aldosterone biosynthesis by carbon monoxide and reversal of this inhibition by light with a maximum effect at a wavelenght of 450 nm. Cytochrome P-450 is a component of mixed function oxidases, which require both molecular oxygen and NADPH, and has been found to be involved in a number of steroid and drug hydroxylation reactions (OMURA et al., 1965).

The conversion of corticosterone to aldosterone and 18-hydroxy-corticosterone was found to be enhanced by bivalent cations; Mg^{++} appeared to be optimal for both reactions (PSYCHOYOS et al., 1966; RAMAN et al., 1966). When Mg^{++} was replaced by Ca^{++} the rate of conversion of corticosterone to 18-hydroxycorticosterone remained equal, but the conversion of corticosterone to aldosterone was markedly inhibited. Both reactions were inhibited by sodium ions (in concentrations above 100 mEq/l) and stimulated by potassium ions (in concentrations above 65 mEq/l) according to SHARMA et al. (1967). However, according to MARUSIC and MULROW (1967b), the conversion of corticosterone to aldosterone by rat adrenal mitochondria was almost the same in potassium phosphate, sodium phosphate, Tris chloride and Krebs-Ringer bicarbonate buffer.

Aldosterone biosynthesis by beef adrenal homogenate was not stimulated by ACTH or angiotensin II (KAHNT and NEHER, 1965).

IV. Factors Directly Influencing Aldosterone Biosynthesis in Short-Term Incubation or Perfusion Experiments

1. Angiotensin II

There is little direct evidence that angiotensin[1] has a stimulating effect on aldosterone biosynthesis by adrenocortical tissue. However, it is a very active stimulator of aldosterone secretion in several animal species and, according to indirect evidence, enhances aldosterone secretion by stimulating biosynthesis rather than by stimulating the release of preformed steroid. Thus, the increments in aldosterone output of isolated dog or sheep adrenals induced by small doses of angiotensin (GANONG et al., 1962; DAVIS, 1962; BLAIR-WEST et al., 1962) were much larger than the total aldosterone content of normal adrenal glands[2]. In the rat, systemically infused angiotensin not only led to a significant increase in aldosterone secretion but also to a sixfold increase in the adrenal content of aldosterone (DUFAU and KLIMAN, 1968a).

In nephrectomized hypophysectomized dogs, injections or infusions of angiotensin into the arterial supply of isolated adrenal glands led within minutes to a distinct elevation of aldosterone, corticosterone and 17-hydroxycorticosteroid secretion (DAVIS, 1962; GANONG et al., 1962). This steroidogenic effect was obtained by very small doses of the peptide (0.5 μg injected or 0.042 μg/min infused), which were ineffective given systemically. In intact sodium-replete sheep with adrenal autotransplants to the neck, infusions of angiotensin II

---

1 Synthetic val[5]-angiotensin II-asp[1] β-amide was used in all investigations referred to in this chapter.

2 A mean content of 0.22 μg of aldosterone per gland was found in five sheep adrenals (COGHLAN and BLAIR-WEST, 1967). The mean aldosterone secretion in sodium-replete sheep of 0.54 μg/h increased severalfold over periods of hours during infusions of angiotensin (BLAIR-WEST et al., 1962).

into the adrenal artery at rates of 0.1 to 0.5 µg/h elicited marked increases in aldosterone secretion without a significant effect on cortisol or corticosterone secretion (BLAIR-WEST et al., 1962, 1970a). Although the aldosterone-stimulating effect of systemically infused angiotensin was first demonstrated in man (LARAGH et al., 1960; BIRON et al., 1961), no evidence has yet been presented that the peptide has a direct effect on the normal human adrenal gland *in vivo* or *in vitro*[3]. Intravenous injection of homologous renin led to a significant increase in aldosterone secretion in the bullfrog (JOHNSTON et al., 1967a) and the opossum (JOHNSTON et al., 1967b). The reports on the effect of exogenous angiotensin on aldosterone secretion in the rat are controversial. Whereas no effect of angiotensin on aldosterone production of normal or hypophysectomized rats was found by EILERS and PETERSON (1964) and MARIEB and MULROW (1965), high doses of the peptide, in excess of those needed to raise blood pressure—3.6 µg/min (DUFAU and KLIMAN, 1968a), 1 µg/min (CADE and PERENICH, 1965)—stimulated aldosterone secretion[4] in intact rats. Aldosterone output rose in response to angiotensin infusions in hypophysectomized nephrectomized rats (MARIEB and MULROW, 1965) and in sodium-deficient nephrectomized hypophysectomized rats (KINSON and SINGER, 1968). It seems unlikely that this lack of activity is due to species differences in the structure of angiotensin, since homologous renin did not stimulate aldosterone secretion in the rat (EILERS and PETERSON, 1964; MARIEB and MULROW, 1965)[5].

In vitro stimulation of aldosterone biosynthesis from endogenous precursors was demonstrated in experiments with beef adrenal tissue. Outer slices of beef adrenal cortex (zona glomerulosa) produced more aldosterone and corticosterone, inner slices (zona fas-

---

3 Angiotensin stimulated aldosterone biosynthesis by incubated tissue slices of adrenocortical adenomas taken from patients with primary aldosteronism but had no effect on contralateral, adjacent or normal adrenocortical tissue (MELBY et al., 1965).

4 In intact rats, the steroidogenic effect of systemically administered angiotensin may be partially due to ACTH release unless ACTH secretion is blocked by glucocorticoids or morphine (VAN DER WAL and DE WIED, 1968; DUFAU and KLIMAN, 1968b).

5 Subcutaneous injections of high amounts of hog renin led to an increased secretion of aldosterone in uninephrectomized rats, but this response became manifest only after a time lag of several hours (MASSON and TRAVIS, 1968).

ciculata) more corticosterone and cortisol when angiotensin (40 µg per gram of tissue) was added to the incubation medium (KAPLAN and BARTTER, 1962). The same effect was observed when lower doses of the peptide (0.1 and 1 µg/g) were used (KAPLAN, 1965). Aldosterone biosynthesis by rat adrenal tissue *in vitro* was not stimulated by angiotensin according to GLASZ and SUGAR (1962), KAPLAN and BARTTER (1962) and SPÄT et al. (1965). However, a small but reproducible increase in aldosterone production was observed when quartered adrenals of sodium-deficient rats were incubated with large amounts of angiotensin (0.3 mg/ml) (MÜLLER, 1965a). In order to compensate for the destruction of angiotensin by adrenal tissue peptidases, the peptide in the medium was periodically replaced (GRÖGER-KORBER, 1966) or was added by constant infusion (SPÄT et al., 1969); however, even under these precautions, angiotensin did not stimulate aldosterone or corticosterone production by incubated rat adrenal tissue. Angiotensin had no effect on corticosteroid production by superfused bisected adrenals of hypophysectomized rats (BANIUKIEWICZ et al., 1968), but, when added in a high concentration to superfused capsular adrenals of hypophysectomized-nephrectomized rats, it stimulated the production of aldosterone, 18-hydroxycorticosterone and corticosterone, although only to a limited extent (TAIT et al., 1970).

A high dose of angiotensin elicited significant steroidogenic responses in incubated capsular adrenals of intact rats, but did not exert a noticeable effect on decapsulated adrenals (MÜLLER, 1971; Fig. 2). It stimulated the production of aldosterone, corticosterone and deoxycorticosterone in capsular adrenals of rats kept on a complete diet, but enhanced only the production of aldosterone in tissue of sodium-deficient rats and only the production of deoxycorticosterone in tissue of potassium-deficient rats.

Since angiotensin stimulates aldosterone, corticosterone and cortisol production by beef or dog adrenals, its action is limited neither to the zona glomerulosa nor to a biosynthetic step specific for the production of aldosterone. However, in low doses angiotensin stimulated only aldosterone secretion in dogs without influencing 17-hydroxycorticosteroid output (GANONG et al., 1966). Similarly, higher doses of angiotensin were necessary to stimulate cortisol biosynthesis by mixed beef adrenocortical tissue than those needed to stimulate aldosterone or corticosterone production (KAP-

LAN, 1965). Thus, angiotensin may have a higher affinity for zona glomerulosa cells in some animal species. On the other hand it apparently acts exclusively on the zona glomerulosa of human, sheep and rat adrenals.

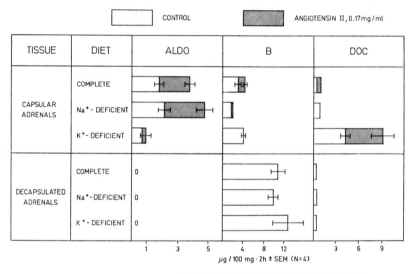

Fig. 2. Effect of angiotensin II (val[5]-angiotensin II-asp[1] β-amide) on *in vitro* production of aldosterone (ALDO), corticosterone (B) and deoxycorticosterone (DOC) by separate zones of adrenals of rats kept on different diets for two weeks. White bars represent base-line corticosteroid production, shaded bars increments due to addition of angiotensin II to the incubation medium. Mean values of two experiments (N = 4) ± standard error of the mean. (Data from MÜLLER, 1971)

All available direct evidence indicates that angiotensin acts on early stages of the aldosterone biosynthetic pathway. Aldosterone production by beef adrenal slices was stimulated by angiotensin in the presence of exogenous cholesterol in substrate amounts but not when progesterone or corticosterone was added to the incubation medium (KAPLAN and BARTTER, 1962). Angiotensin did not stimulate the conversion of tritiated corticosterone to aldosterone while it stimulated the production of unlabelled aldosterone from endogenous precursors in sheep adrenal perfusion studies (BLAIR-WEST et al., 1967; BANIUKIEWICZ et al., 1968). In rat adrenal tissue angiotensin did

Table 1. *Incubation of rat adrenal quarters with trace amounts of tritiated precursors. Mean values of two flasks ± range.* (From MÜLLER, 1966)

Substrate	Addition to incubation medium	Conversion to aldosterone-³H cpm/100mg ×10⁻³	Conversion to corticosterone-³H cpm/100mg ×10⁻³	Production of corticosterone[a] μg/100 mg
Corticosterone-1, 2-³H (945 600 cpm per 100 mg tissue)	—	16.2 ± 0.9		9.1 ± 0.6
	NH$_4$Cl[b]	16.7 ± 3.0		9.0 ± 1.2
	NADP+Glc-6-P[c]	20.6 ± 1.4		63.7 ± 3.7
	ACTH[d]	13.6 ± 3.7		92.3 ± 6.5
	—	21.9 ± 0.1		13.3 ± 0.4
	KCl[e]	25.7 ± 1.4		16.1 ± 0.7
	Angiotensin II[f]	17.4 ± 0.7		13.1 ± 0.4
Pregnenolone-7α-³H (1 505 000 cpm per 100 mg tissue)	—	69.6 ± 3.2	393 ± 3	10.8 ± 0.7
	NH$_4$Cl	66.6 ± 3.5	370 ± 22	16.6 ± 0.4
	NADP+Glc-6-P	74.0 ± 9.8	637 ± 60	71.4 ± 3.2
	ACTH	60.8 ± 3.7	390 ± 8	93.6 ± 0.4
	—	71.2 ± 3.0	318 ± 6	14.7 ± 0.6
	KCl	73.2 ± 9.4	334 ± 13	18.2 ± 0.9
	Angiotensin II	74.1 ± 4.1	302 ± 17	15.4 ± 0.9
Cholesterol-7α-³H (21.4×10⁶ cpm per 100 mg tissue, added in Tween 80)	—	1.34 ± 0.11	3.40 ± 0.08	7.7 ± 0.3
	NH$_4$Cl	2.40 ± 0.26	4.37 ± 0.35	9.0 ± 0.5
	NADP+Glc-6-P	27.64 ± 1.74	197.53 ± 5.96	48.5 ± 0.6
	ACTH	3.94 ± 0.30	15.25 ± 1.80	64.3 ± 5.2
	—	0.36 ± 0.03	2.54 ± 0.89	11.2 ± 1.3
	KCl	3.26 ± 0.80	4.82 ± 0.92	14.7 ± 1.3
	Angiotensin II	0.80 ± 0.02	2.44 ± 0.16	11.6 ± 0.1

[a] fluorimetric assay. [b] 7.7 mmol/l. [c] NADP-Na + glucose-6-phosphate-Na$_2$; 10.5 μmol/flask. [d] 5 IU/flask. [e] 8.1 mEq K$^+$/l. [f] Val^5-angiotensin II -asp^1 β-amide; 2 mg/flask.

not increase the incorporation of tritiated pregnenolone, progesterone, deoxycorticosterone or corticosterone into aldosterone, whereas it stimulated the conversion of tritiated cholesterol to aldosterone (MÜLLER, 1966, Table 1). However, in beef adrenal tissue it stimulated the conversion of ^{14}C-labelled acetate, but not that of ^{14}C-cholesterol or ^{14}C-progesterone, to aldosterone and other corticosteroids (LOMMER and WOLFF, 1966). Thus, while there is agreement on an early site of action of angiotensin on aldosterone biosynthesis, preceding the formation of progesterone and thereby probably excluding an effect on 18-hydroxylation and 18-hydroxydehydrogenation, present evidence does not allow firm conclusions as to the exact step in the biosynthetic pathway which is influenced by angiotensin.

2. Monovalent Cations

a) Sodium

A direct influence of the ionic environment on aldosterone biosynthesis was first observed by ROSENFELD et al. (1956), who found a significant increase in the secretion of aldosterone-like material by isolated calf adrenals when the sodium concentration of the perfusing fluid was decreased from 149 to 125 mEq/l and the potassium concentration was raised from 3.5 to 25 mEq/l. DENTON et al. (1959) demonstrated that increases of potassium and decreases of sodium concentrations within physiological ranges in the adrenal arterial blood of sheep with adrenal autotransplants to the neck led to changes in the electrolyte composition of parotid saliva indicative of an increased secretion of aldosterone. Further studies by this group (BLAIR-WEST et al., 1962) confirmed that local alterations of sodium and potassium concentration in the adrenal arterial blood significantly influenced aldosterone secretion—directly measured by a double isotope dilution derivative assay. Although a simultaneous increase in potassium and decrease in sodium concentration was the most effective stimulus, independent changes in the concentration of either cation could influence aldosterone production (BLAIR-WEST et al., 1963). A decrease in sodium concentration by 5 to 20 mEq/l significantly stimulated aldosterone secretion in sodium-replete animals, without having a consistent effect on corticosterone or cortisol output. In hypophysectomized nephrectomized dogs, infusion of 5% glucose solution to the arterial supply of isolated adrenals, which lowered local plasma sodium concentration by 20m Eq/l, significantly increased aldosterone secretion (DAVIS et al., 1963).

In sheep with established sodium deficiency, in which plasma sodium concentration was 9 mEq/l lower than in sodium-replete sheep, a local increase in the adrenal arterial plasma sodium level of 8 mEq/l lowered aldosterone secretion by 25% for periods of one to three hours (BLAIR-WEST et al., 1966). However, aldosterone secretion returned to its initial high rate after 3 to 6 hours, even when the local sodium concentration was increased by 15 mEq/l. Infusions of hypertonic mannitol into the adrenal artery had no effect on aldosterone secretion. At the onset of sodium deficiency the effects of high local sodium concentrations were more pronounced.

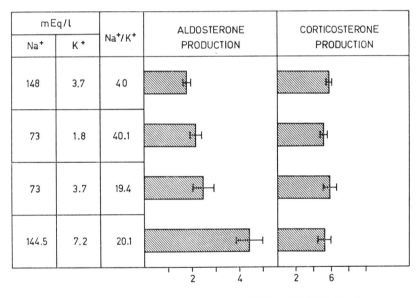

μg / 100 mg · 2 h ± SD (N=4)

Fig. 3. Effect of sodium and potassium concentration in the incubation medium (mixture of Krebs-Ringer bicarbonate buffer, isotonic glucose solution, and 0.154 M KCl) on aldosterone and corticosterone production by rat adrenal quarters *in vitro*. Mean values of two experiments (N = 4) ± standard deviation. (For methodology see MÜLLER, 1965a)

Although the normalisation of the small decrease in plasma sodium of 2—4 mEq was not sufficient to stop the rapid increase in the aldosterone secretion rate, an elevation of adrenal plasma sodium concentration by 6 to 15 mEq.l caused marked and sustained decreases in aldosterone production. In sodium-replete sheep, the aldosterone-stimulating effect of angiotensin II was significantly reduced when the peptide was infused in a hypertonic saline solution causing a mean increase of 10 mEq/l in the adrenal arterial plasma sodium level (BLAIR-WEST et al., 1965a).

Whereas relatively small changes in the potassium concentration of the incubation medium significantly influence aldosterone production by rat and beef adrenal tissue *in vitro* (KAPLAN, 1965; MÜLLER, 1965a), even marked decreases in sodium concentration are either ineffective or lead only to small increments in aldosterone production (Fig. 3). Thus a decrease of sodium concentration from 148 to 73 mEq/l resulted only in a 40% increase in aldosterone production, whereas in the same tissue a 150% stimulation of the aldosterone output was observed when the potassium concentration was increased from 3.7 to 7.2 mEq/l, although the sodium/potassium ratio was in both instances reduced from 40 to 20.

b) Potassium

Small increases in the potassium concentration of the adrenal arterial plasma—with or without concomitant decreases in sodium concentration—can significantly stimulate aldosterone secretion, as was first shown in studies with sheep by BLAIR-WEST et al. (1962, 1963, 1970a). An elevation of the potassium concentration by 0.5 mEq/l or less was a potent stimulus of aldosterone production but had no effect on corticosterone or cortisol output (FUNDER et al., 1969). In hypophysectomized nephrectomized dogs, infusions of KCl or K_2SO_4 solution into the arterial supply of isolated adrenal glands repeatedly stimulated aldosterone secretion and occasionally also corticosterone secretion (DAVIS et al., 1963). Systemic infusion of potassium chloride in hypophysectomized dogs augmented the secretion rate of both aldosterone and corticosterone. The smallest increment in plasma potassium concentration eliciting a steroidogenic response was 1.3 mEq/l.

A direct effect of potassium ions in the incubation medium on aldosterone production by rat adrenal tissue *in vitro* was first demonstrated by GIROUD et al. (1956b); aldosterone production was 40% higher in a medium with a potassium concentration of 8.6 mEq/l than in a medium with a potassium concentration of 5.8 mEq/l. Aldosterone and corticosterone production by beef adrenal slices rose proportionally to the potassium concentration in the medium over a range between 0 and 14.5 mEq/l (KAPLAN, 1965). When quartered adrenal glands of sodium-deficient rats were incubated in buffers with different potassium concentrations, aldosterone production rose with increasing K^+ levels to reach a maximum between 7 and 9 mEq K^+/l (MÜLLER, 1965a, Fig. 4). If potassium concentration was elevated to values above 10 mEq/l, there was a significant decrease in aldosterone production. Whereas corticosterone production was not significantly affected by the potassium concentration, the production of deoxycorticosterone was stimulated by a high potassium level (MÜLLER, 1968). The steroidogenic response of rat adrenal tissue *in vitro* to potassium ions was found to

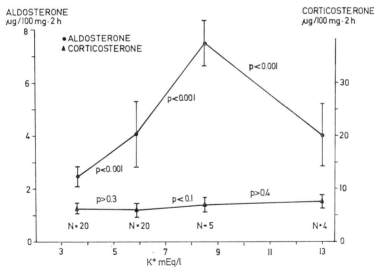

Fig. 4. Effect of potassium concentration in the incubation medium on *in vitro* production of aldosterone and corticosterone by rat adrenal quarters. Mean values ± standard deviaton. P values of significance were calculated by *t* tests. (From MÜLLER, 1965a)

2*

depend to a considerable extent on sodium and potassium balance *in vivo*. Potassium ions stimulated deoxycorticosterone production to a greater and aldosterone production to a lesser extent in adrenal tissue of rats which had received sodium chloride solution as drinking fluid (MÜLLER, 1968). In adrenals of potassium-deficient rats a high potassium concentration *in vitro* had no effect on aldosterone production but elicited a significant increase in deoxycorticosterone production (MÜLLER and HUBER, 1969; MÜLLER, 1971, Fig. 5). Potassium ions stimulated aldosterone more and deoxycorticosterone

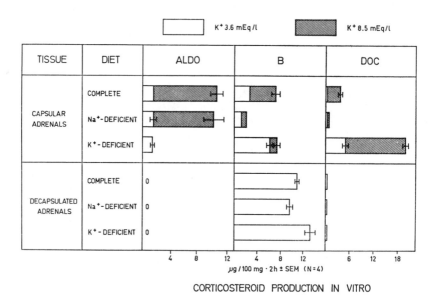

CORTICOSTEROID PRODUCTION IN VITRO

Fig. 5. Production of aldosterone (ALDO), corticosterone (B), and deoxycorticosterone (DOC) by separate zones of adrenals of rats kept on different diets for two weeks. White bars represent base-line corticosteroid production, shaded bars increments due to addition of KCl to the incubation medium. Mean values of two experiments (N = 4) ± standard error of the mean. (Data from MÜLLER, 1971)

production less markedly in adrenal tissue of rats which had been kept on a sodium-deficient diet for two weeks or were uremic 40 hours after bilateral nephrectomy.

During short-term incubation, of rat adrenal tissue potassium ions act on early stages of the aldosterone biosynthetic pathway

Table 2. *Effect of serotonin and potassium ions on corticosteroid production by rat adrenal quarters. Mean values of two flasks ±range.* (For methodology see MÜLLER, 1965a; MÜLLER and WEICK, 1967)

mEq K+/l in incubation medium	Corticosteroid production µg/100 mg/2 h					
	Aldosterone		Corticosterone		Deoxycorticosterone	
	Control	Serotonin[a]	Control	Serotonin	Control	Serotonin
4.0	3.72±0.02	7.19±0.16	7.00±0.94	6.41±0.20	0.39±0.10	0.39±0.02
0	3.01±0.23	4.75±0.02	5.91±0.46	5.44±0.08	0.35±0.04	0.42±0.02
3.6	1.42±0.16	5.08±0.83	7.02±0.08	8.90±0.44	0.14±0.01	0.77±0.04
8.5	4.19±0.97	4.64±0.73	9.28±0.92	9.80±0.39	0.92±0.26	0.87±0.21

[a] 5-hydroxytryptamine creatinine sulfate, 1.7×10^{-5} M.

and influence mainly the conversion of cholesterol to pregnenolone according to studies with tritiated precursor steroids (MÜLLER, 1966, Table 1). Only the incorporation of tritiated cholesterol into aldosterone was stimulated by a high potassium concentration in the incubation fluid; incorporation of tritiated pregnenolone, progesterone, deoxycorticosterone and corticosterone, respectively, into aldosterone was the same in a low potassium (3.6 mEq/l) and a high potassium (8.1 mEq/l) medium. On the other hand, evidence presented by BURWELL et al. (1969) indicates that potassium ions stimulate steroidogenesis in dog adrenal slices by enhancing 11β-hydroxylation.

Potassium ions exert their steroidogenic activity exclusively on the zona glomerulosa of the rat adrenal cortex. They stimulated the production of aldosterone, corticosterone and deoxycorticosterone in the capsular portion of rat adrenal glands (containing the zona glomerulosa) and did not influence corticosteroid biosynthesis in the decapsulated portion (containing zona fasciculata and zona reticularis) (MÜLLER, 1971, Fig. 5). In beef adrenal tissue, potassium ions enhanced aldosterone and corticosterone output but not the production of cortisol, which derives from the zona fasciculata (KAPLAN, 1965). By contrast, potassium ions also stimulated cortisol production by outer slices of dog adrenals (BURWELL et al., 1969).

In beef adrenal tissue potassium ions potentiated the aldosterone-stimulating effect of angiotensin (KAPLAN, 1965). A high potassium concentration did not further stimulate steroidogenesis in rat adrenal tissue maximally stimulated by serotonin (Table 2). Thus, potassium and serotonin may act on the same cells. Stimulation of aldosterone production by potassium ions can be blocked by ouabain (Table 9), cycloheximide (Table 8) and the absence of calcium ions in the incubation medium (Table 3). In a potassium-free medium the aldosterone-stimulating effect of serotonin was diminished but not abolished (Table 2).

c) Ammonium, Rubidium, and Caesium

Addition of ammonium chloride or ammonium hydroxide to the incubation medium significantly stimulated aldosterone production *in vitro* by adrenal tissue of sodium-deficient rats and had only

negligible effects on corticosterone production (MÜLLER, 1965b).
Rubidium chloride or caesium chloride had a similar effect on corticosteroid biosynthesis (MÜLLER, 1965c). When increasing amounts
of these cations were added to the incubation medium, aldosterone production was highest at concentrations of approximately 7 mEq/l and
was significantly lower at higher concentrations (Fig. 6); a similar

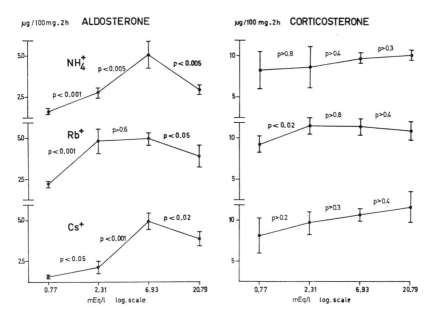

Fig. 6. Effect of increasing concentrations of NH₄Cl, RbCl, and CsCl in
the incubation medium on *in vitro* production of aldosterone and corticosterone by rat adrenal quarters. Mean values of two experiments (N = 4)
± standard deviation. *P* values of significance were calculated by *t* tests.
(From MÜLLER, 1965c)

dose/response curve had been observed with increasing potassium
concentrations (Fig. 4). The effects of K^+ and NH_4^+ ions were addi-
tive only at low concentrations. Adrenal tissue maximally stimulated by potassium did not further respond to ammonium ions.

Aldosterone and corticosterone production *in vivo* by isolated dog
adrenals rose in response to local infusion of CsCl or RbCl solution
(BARTTER et al., 1964). On the other hand, rubidium had no aldosterone-stimulating effect on perfused adrenals of hypophysectomized dogs according to HILTON et al. (1965). Ammonium chloride

in concentrations greater than 1 mEq/l stimulated aldosterone secretion by perfused sheep adrenals to levels seen in sodium-deficient sheep (BLAIR-WEST et al., 1968d). Ammonium was found to stimulate aldosterone biosynthesis at the level of the conversion of cholesterol to pregnenolone and to have no effect on later biosynthetic steps (MÜLLER, 1966, Table 1).

3. Bivalent Cations

a) Calcium

ACTH stimulates the corticosteroid biosynthesis in adrenal tissue *in vitro* only when the incubation medium contains calcium (BIRMINGHAM et al., 1953; PERON and KORITZ, 1958). Similarly the presence of calcium ions is required for the steroidogenic effect of certain aldosterone-stimulating agents. The *in vitro* response in aldosterone production by rat adrenal tissue to stimulation by potassium ions or angiotensin II was completely abolished in a calcium-free medium, whereas the absence of calcium had no consistent effect on base-line aldosterone or corticosterone production (Table 3).

Short-term infusions of calcium chloride into the adrenal arterial blood supply of conscious sheep with autotransplanted adrenal glands leading to increases in local calcium concentration by up to 9.5 mEq/l did not influence aldosterone secretion (BLAIR-WEST et al., 1968d).

b) Magnesium

When rat adrenal tissue was preincubated in a calcium- and magnesium-free medium containing EDTA, addition of calcium only to the medium used for the final incubation restored the response in aldosterone production to stimulation by a high potassium concentration, whereas the presence of magnesium ions was not required (Table 3).

In sheep, adrenal arterial infusions of magnesium chloride which elevated the local plasma magnesium concentration by up to 5.5 mEq/l had no effect on aldosterone secretion (BLAIR-WEST et al., 1968d).

Table 3. *Effect of calcium and magnesium ions on aldosterone and corticosterone production by rat adrenal quarters in vitro. Mean values of two flasks ±range.* (For methodology see MÜLLER, 1965a)

Preincubation medium[a] (mEq/l)			Incubation medium[a] (mEq/l)		Addition	Steroid production (μg/100 mg/2 h)	
EDTA[b]	Ca++	Mg++	Ca++	Mg++		Aldosterone	Corticosterone
+	0	0	2.6	2.4	—	0.62±0.12	7.15±0.73
+	0	0	2.6	2.4	KCl[c]	2.48±0.25	8.79±0.19
+	0	0	0	2.4	—	0.66±0.03	6.51±0.01
+	0	0	0	2.4	KCl	0.69±0.07	6.92±0.06
—	2.6	2.4	2.6	2.4	—	2.05±0.32	6.27±0.11
—	2.6	2.4	2.6	2.4	Angiotensin II[d]	4.46±0.22	6.78±0.40
+	0	0	0	2.4	—	1.13±0.06	3.54±0.13
+	0	0	0	2.4	Angiotensin II	1.20±0.01	3.90±0.16
+	0	0	2.6	2.4	—	1.05±0.23	6.66±0.23
+	0	0	2.6	2.4	KCl	2.27±0.11	8.88±0.04
+	0	0	2.6	0	—	0.92±0.12	7.33±0.21
+	0	0	2.6	0	KCl	2.29±0.09	9.89±0.42

a 3.6 mEq K+/l. b Dihydrate of disodium ethylenediamine tetraacetate, 10^{-3} M. EDTA was not added to the medium used for the final incubation in these studies. c 8.5 mEq K+/l. d Val[5]-angiotensin II-asp[1] β-amide, 0.17 mg/ml.

Whereas magnesium ions apparently do not directly influence aldosterone biosynthesis in short-term experiments, prolonged dietary magnesium deficiency in rats has been found to induce considerable increases in aldosterone secretion (GINN et al., 1967) in the zona glomerulosa width and in the juxtaglomerular granulation index of the kidneys (CANTIN and VEILLEUX, 1968), as well as decreases in the peripheral plasma concentration and the adrenal content of corticosterone (RICHTER et al., 1968). The mechanism by which magnesium balance can mediate alterations in adrenocortical function and in the renin-angiotensin system is unknown.

4. ACTH

Whereas the biosynthesis and secretion of glucocorticosteroids is almost exclusively controlled by ACTH (adrenocorticotrophic hormone), the production of aldosterone is relatively independent of ACTH. Increased secretion of cortisol or corticosterone due to chronic excessive secretion of endogenous ACTH or to long-term medication with exogenous ACTH may be accompanied by unchanged or diminished aldosterone secretion. However, in short-term experiments, ACTH has been consistently found to stimulate aldosterone production as well as glucocorticosteroid production. Aldosterone stimulation by ACTH must always be considered in experiments which lead to surgical stress and which are carried out in animals with intact pituitary glands. ACTH stimulated aldosterone production *in vitro* by adrenal tissue of man (DYRENFURTH et al., 1960), ox (KAPLAN and BARTTER, 1962), rat (GIROUD et al., 1956; LUCIS et al., 1961; KAPLAN and BARTTER, 1962; MÜLLER, 1965a, Fig. 7), duck (DONALDSON and HOLMES, 1965), turtle (MACCHI, 1963) and bullfrog (CARSTENSEN et al., 1961) and by perfused calf (ROSENFELD et al., 1956) and dog adrenals (GREENWAY and VERNEY, 1962) as well as by a hyperplastic adrenal gland surgically removed from a patient with Cushing's disease (SCHRIEFERS et al., 1963). ACTH enhanced aldosterone secretion *in vivo* (or urinary excretion of aldosterone-18-glucuronide) in man (VENNING et al., 1956; MULLER et al., 1956; CRABBE et al., 1959; TUCCI et al., 1967), dog (FARRELL et al., 1955; HILTON et al., 1960; MULROW and GANONG, 1961), sheep (BLAIR-WEST et al., 1962), rat (SINGER and STACK-

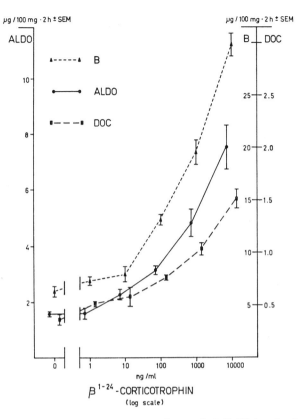

Fig. 7. Effect of increasing concentrations of ACTH in the incubation medium on *in vitro* production of corticosterone (B), aldosterone (ALDO), and deoxycorticosterone (DOC) by rat adrenal quarters. Mean values of three experiments (N = 4) ± standard error of the mean. (For methodology see MÜLLER, 1965a; MÜLLER and WEICK, 1967)

DUNNE, 1955; KINSON and SINGER, 1968), opossum (JOHNSTON et al., 1967b) and bullfrog (JOHNSTON et al., 1967a; ULICK and FEINHOLTZ, 1968). In most of these experiments the fractional increase in aldosterone output was smaller than that observed in glucocorticosteroid output. In nephrectomized hypophysectomized dogs, aldosterone production was only stimulated by a dose of ACTH tenfold that necessary for maximal glucocorticosteroid stimulation (GANONG et al., 1965). Similarly, in stressed dogs, in which glucocorticosteroid production was already maximally stimulated by endogenous ACTH, a pharmacological dose of exogenous ACTH resulted in an increase of aldosterone production only (MULLER et al., 1964).

The response in aldosterone production to ACTH can be experimentally modified by changes in sodium and potassium balance or by treatment with renin. ACTH elicited a more marked or longer lasting increase in aldosterone secretion in men on a sodium-deficient diet than in persons with a normal or high sodium intake (MULLER et al., 1956; LIDDLE et al., 1956; VENNING et al., 1962; TUCCI et al., 1967). In dogs kept on a sodium-deficient diet for five days prior to hypophysectomy and nephrectomy, ACTH stimulated aldosterone secretion at a lower dosage and to a greater extent than in sodium-replete animals (GANONG et al., 1965). Longer periods of sodium restriction further enhanced the aldosterone-stimulating effect of ACTH. On the other hand, sodium deficiency did not influence the response in 17-hydroxycorticosteroid output. The aldosterone-stimulating effect of ACTH was also enhanced in dogs which had been injected for five days with dog renin (GANONG et al., 1967). Injection of ACTH in hypophysectomized rats *in vivo* induced significant increases in aldosterone production by adrenal tissue *in vitro* only when the animals were kept on a sodium-deficient diet (CSANKY et al., 1968). However, KINSON and SINGER (1968) found that the effect of ACTH on aldosterone secretion *in vivo* in hypophysectomized nephrectomized rats was the same in animals kept on a normal or a sodium-deficient diet. ACTH stimulated aldosterone production *in vitro* more markedly in adrenal tissue of rats that were kept on a sodium-deficient diet (MÜLLER, 1965a)[6] or were uremic after bilateral nephrectomy (MÜLLER and HUBER, 1969) than in tissue of respective control animals. On the other hand, ACTH stimulated aldosterone production to a lesser degree in adrenals of rats receiving sodium chloride solution as drinking fluid

---

6 During incubation with added ACTH, aldosterone production was the same in capsular adrenals of sodium-replete and sodium-deficient rats (MÜLLER, 1971; Fig. 8). Therefore, the increased response to ACTH in aldosterone production by quartered whole adrenals of sodium-deficient rats may have been due to increased conversion to aldosterone of corticosterone which had been produced by the zona fasciculata under the influence of ACTH and which entered the zona glomerulosa cells either by direct intercellular diffusion or from the incubation medium. Thus, the apparent increase in responsiveness of the zona glomerulosa to ACTH stimulation may have been an artefact of this particular *in vitro* system.

(Müller, 1968) and did not stimulate aldosterone production in adrenals of potassium-deficient rats (Müller and Huber, 1969).

ACTH acted mainly on the zona fasciculata of beef adrenals and had only a negligible steroidogenic effect on the zona glomerulosa, according to Stachenko and Giroud (1959). However, according to Kaplan (1965), it stimulated aldosterone and corticosterone production by outer slices of beef adrenals although not to the same extent it stimulated cortisol production by the inner zones. Thus, ACTH may have a lower affinity to zona glomerulosa cells. On the other hand, at least high doses of ACTH were capable of stimulating steroidogenesis severalfold in superfused capsular adrenals of intact and particularly of hypophysectomized rats (Baniukiewicz et al., 1968; Tait et al., 1970). Maximum stimulation of corticosteroid production by capsular adrenals of normal, sodium-deficient and potassium-deficient rats by high doses of ACTH was also observed in our laboratory (Müller, 1971; Fig. 8).

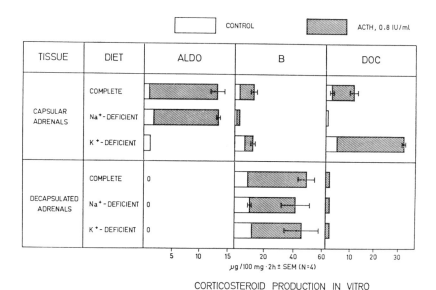

Fig. 8. Production of aldosterone (ALDO), corticosterone (B), and deoxycorticosterone (DOC) by separate zones of adrenals of rats kept on different diets for two weeks. White bars represent base-line corticosteroid production, shaded bars increments due to addition of ACTH (0.8 IU/ml) to the incubation medium. Mean values of two experiments (N = 4) ± standard error of the mean. (Data from Müller, 1971)

During short-term perfusion and incubation studies, the main site of action of ACTH on the biosynthesis of glucocorticosteroids is the conversion of cholesterol to pregnenolone (STONE and HECHTER, 1954; KARABOYAS and KORITZ, 1965). In the aldosterone biosynthetic pathway, also, ACTH acts only on the conversion of cholesterol to pregnenolone and does not influence the reactions between pregnenolone and aldosterone (Table 1). Thus, ACTH stimulated the incorporation of radioactively labelled cholesterol, but not of labelled pregnenolone, progesterone, deoxycorticosterone or corticosterone, into aldosterone by rat or beef adrenal tissue (MÜLLER, 1966; LOMMER and WOLFF, 1966a). ACTH did not stimulate aldosterone production by beef adrenal slices when unlabelled corticosterone, deoxycorticosterone or progesterone were added in substrate amounts to the incubation medium (KAPLAN and BARTTER, 1962).

5. Cyclic AMP

For a decade a considerable amount of research on the biochemical mechanism of action of ACTH on the adrenal cortex has been based on the theory proposed by HAYNES and BERTHET (1957)[7]. Although present evidence does not confirm this theory *in toto* (HILF, 1965; HALKERSTON, 1968), it has been amply demonstrated that cyclic AMP (cyclic adenosine-3',5'-monophosphate) imitates the steroidogenic and the adrenal weight-maintaining effect of ACTH *in vitro* and *in vivo* (HAYNES et al., 1959; HILTON et al., 1961; IMURA et al., 1965; NEY, 1969). Furthermore, ACTH leads to dose-dependent increases in adrenocortical cyclic AMP-concentrations, which are maintained as long as steroidogenesis remains elevated (HAYNES, 1958; GRAHAME-SMITH et al., 1967). Cyclic AMP stimulated aldosterone production *in vitro* by beef adrenal slices in a similar manner as ACTH (KAPLAN, 1965)

---

7 According to this theory, ACTH stimulates the formation of cyclic AMP. Cyclic AMP activates phoshporylase, which in turn mediates the conversion of glycogen to glucose-l-phosphate. Glucose-l-phosphate is converted to glucose-6-phosphate by phosphoglucomutase. Metabolism of glucose-6-phosphate through the hexose monophosphate shunt results in the production of NADPH, a coenzyme necessary in several steps of steroidogenesis.

It had no cumulative effect when added together with ACTH, but it potentiated the aldosterone-stimulating effect of angiotensin. Like ACTH, cyclic AMP stimulated both aldosterone and corticosterone production by rat adrenal tissue and stimulated steroidogenesis in capsular (zona glomerulosa) as well as in decapsulated (zona fasciculata-reticularis) glands (MÜLLER, 1971, Table 4).

Table 4. *Effect of dibutyryl cyclic AMP on corticosteroid production by adrenal tissue of sodium-deficient rats. Mean values of two experiments (N = 4) ± standard deviation in µg/100 mg/2 h. p Values of significance were calculated by t tests and refer to differences between adjacent lines* (From MÜLLER, 1971)

Tissue	Addition to incubation medium	Steroid Production Aldosterone	Corticosterone
Capsular adrenals	—	1.71±0.42 ($p < 0.001$)	1.84±0.44 ($p < 0.001$)
	cyclic AMP[a]	11.31±2.44	5.42±0.35
Decapsulated adrenals	—		11.79± 2.67 ($p < 0.001$)
	cyclic AMP[a]		50.72±11.17

[a] Dibutyryl adenosine-3',5'-monophosphate, 10^{-3} M

6. NADPH

NADP and glucose-6-phosphate, added to the medium in which quartered rat adrenals were incubated, elicited an increase in corticosteroid production similar to the one induced by ACTH (KORITZ and PERON, 1958). However, the steroidogenic effects of ACTH and NADPH were found to be cumulative in this *in vitro* preparation, possibly because ACTH acted only on intact cells, exogenous NADPH only on damaged cells (HALKERSTON et al. 1968). NADP and glucose-6-phosphate stimulated aldosterone and corticosterone production from endogenous precursors in rat adrenal tissue (MÜLLER, 1965c), but did not affect the conversion of tritiated corticosterone, deoxycorticosterone, progesterone and pregnenolone to aldosterone

(MÜLLER, 1966, Table 1). NADPH stimulated the conversion of ^{3}H-cholesterol into aldosterone but appeared to act on a cholesterol pool, which was different from the one utilized for steroid biosynthesis under stimulation by other aldosterone-stimulating substances. It stimulated the incorporation of tritiated cholesterol into aldosterone approximately twentyfold and the production of unlabelled aldosterone from endogenous precursors only threefold, whereas ACTH, potassium ions, ammonium ions, angiotensin and serotonin, respectively, led to the same fractional increases in aldosterone production and in the incorporation of tritiated cholesterol into aldosterone.

7. Serotonin

The steroidogenic effect of serotonin (5-hydroxytryptamine[8]) on adrenocortical tissue was discovered by Rosenkrantz and co-workers (ROSENKRANTZ, 1959; ROSENKRANTZ and LAFERTE, 1960; CONNORS and ROSENKRANTZ, 1962). Small amounts of serotonin stimulated the *in vitro* production of a blue tetrazolium reducing steroid, which was tentatively identified as aldosterone by its chromatographic properties and by its marked sodium-retaining biological activity. This corticosteroid-stimulating effect was observed in incubates of rabbit, guinea pig, rat and cow adrenal tissue and of slices of a human adrenocortical tumor. Between 30 and 50% of the added serotonin was taken up by the adrenal tissue during 90 minutes of incubation. Serotonin also stimulated oxygen consumption by the adrenal tissue. Its steroidogenic activity was abolished by the absence of oxygen or by the presence of cyanide or dinitrophenol. Corticosteroid production by cow adrenocortical mitochondria incubated in phosphate buffer and in an atmosphere of oxygen was also enhanced by serotonin; this effect was potentiated by ADP or ATP but not by cyclic AMP. Jouan and Samperez (JOUAN, 1963, 1967; JOUAN and SAMREPEZ, 1965) found that serotonin had a steroidogenic effect on rat adrenal tissue incubates which was similar to that of pineal gland extracts and different from that of ACTH.

---

8 5-hydroxytryptamine creatinine sulphate was used in all experiments referred to in this chapter.

Serotonin and pineal gland extracts stimulated mainly aldosterone and 18-hydroxycorticosterone production and had only a small effect on corticosterone and 18-hydroxy-11-deoxycorticosterone output; the main effect of ACTH was on the production of the latter two corticosteroid fractions. This specific aldosterone-stimulating effect of serotonin on rat adrenal incubates was confirmed by us (MÜLLER and ZIEGLER, 1968) when we isolated aldosterone-stimulating material from rat serum dialysate (MÜLLER and WEICK, 1967) and found evidence that this activity was due to the presence of serotonin.

On the other hand, there is little evidence that serotonin has a comparable corticosteroid-stimulating acitivty *in vivo*. Serotonin stimulated the secretion of cortisol and corticosterone by perfused adrenals of hypophysectomized dogs to the same extent as ACTH (VERDESCA et al., 1961). A high aldosterone output was measured under serotonin stimulation; however, aldosterone production was not determined during the control period. Farrell and McIsaac found serotonin a hundred times less active than "adrenoglomerulotropin" (6-methoxytetrahydroharman[9]) in stimulating aldosterone secretion in the mid-collicular brain-removed, hypophysectomized dog (FARRELL and MCISAAC, 1961). Serotonin did not stimulate aldosterone, corticosterone or cortisol secretion in sodium-replete sheep (BLAIR-WEST et al., 1962).

In vitro serotonin appears to be one of the most active aldosterone-stimulating substances. It significantly stimulated aldosterone production by quartered adrenals from rats kept on a sodium-deficient diet for two weeks in a concentration as low as 10^{-8} mol/l (MÜLLER and ZIEGLER, 1968; Fig. 9). The log-dose/response curve showed an approximately linear slope between serotonin concentrations of 10^{-8} and 10^{-6} mol/l and a plateau between concentrations of 10^{-6} and 10^{-4} mol/l. A similar but smaller effect was seen on deoxycorticosterone production, whereas corticosterone production was not stimulated by any dose of serotonin. The steroidogenic response of rat adrenal tissue to serotonin was to a large extent dependent on the sodium and potassium balance of the animals from which the adrenals were taken (MÜLLER and HUBER, 1969; Fig. 10). Serotonin

---

9 This substance was found to have no aldosterone-stimulating activity either in the sheep (BLAIR-WEST et al. 1963) and dog (MULROW et al. 1963) *in vivo* nor on rat adrenal tissue *in vitro* (JOUAN, 1967; MÜLLER and ZIEGLER, 1968).

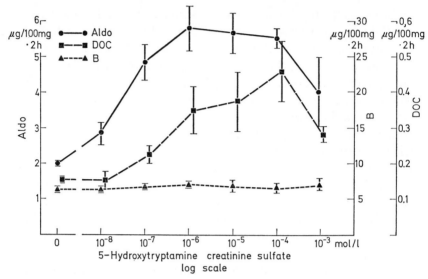

Fig. 9. Effect of increasing concentrations of serotonin (5-hydroxytrypta-
mine creatinine sulfate) in the incubation medium on *in vitro* production of
aldosterone (ALDO), corticosterone (B), and deoxycorticosterone (DOC)
by rat adrenal quarters. Mean values of four experiments (N = 4) ± standard error of the mean. (From MÜLLER and ZIEGLER, 1968)

Table 5. *Incorporation of tritiated precursor steroids into aldosterone and corticosterone by quartered rat adrenals. Mean values of two flasks ± range.* (From
MÜLLER and ZIEGLER, 1968)

Substrate	Addition to incubation medium	Conversion to	
		aldoste-rone-^{3}H cpm/100mg $\times 10^{-3}$	cortico-sterone-^{3}H cpm/100mg $\times 10^{-3}$
Cholesterol-7α ^{3}H 74.5 × 10⁶ cpm	—	2.23 ± 0.03	12.9 ± 0.5
per 100 mg tissue	Serotonin (10⁻⁵ M)	5.14 ± 0.35	16.8 ± 0.8
Pregnenolone-7α ^{3}H 1.31 × 10⁶ cpm	—	87.5 ± 4.2	369 ± 11
per 100 mg tissue	Serotonin (10⁻⁵ M)	104.5 ± 13.5	349 ± 12

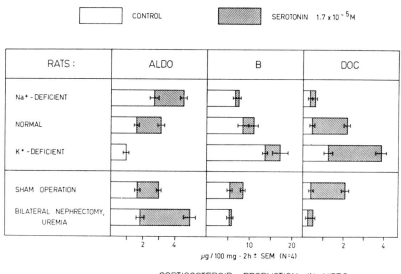

CORTICOSTEROID   PRODUCTION   IN   VITRO

Fig. 10. Production of aldosterone (ALDO), corticosterone (B), and deoxycorticosterone (DOC) by adrenal tissue of rats kept on different diets for
two weeks and rats uremic 40 hours after bilateral nephrectomy. White
bars represent base-line corticosteroid production, shaded bars increments due to addition of serotonin (5-hydroxytryptamine creatinine sulfate,
1.7×10^{-5}M) to the incubation medium. Mean values of two experiments
(N = 4) $\pm$ standard error of the mean. (Data from MÜLLER and HUBER,
1969)

stimulated aldosterone production in adrenals of sodium-deficient
and normal rats, but not in tissue of potassium-deficient rats, where
it stimulated DOC production more markedly. In adrenals of rats
which were uremic 40 hours following bilateral nephrectomy, serotonin stimulated aldosterone production to a greater and deoxycorticosterone production to a smaller extent than in adrenals of
sham-operated animals.

Serotonin stimulated the incorporation of tritiated cholesterol
into aldosterone by 130% and the incorporation into corticosterone
by 30% (MÜLLER and ZIEGLER, 1968, Table 5); it enhanced the conversion of tritiated pregnenolone to aldosterone only slightly (+19%)
and did not influence the conversion to corticosterone. This indicated that serotonin acted mainly on the conversion of cholesterol
to pregnenolone and had only a small effect on the later steps of bio-

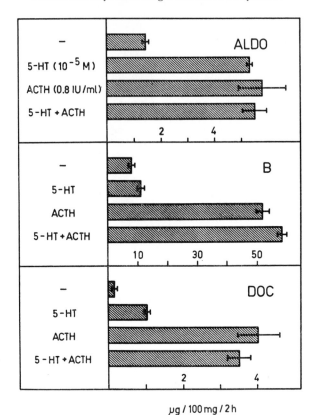

Fig. 11. Production of aldosterone (ALDO), corticosterone (B), and deoxy-corticosterone (DOC) by rat adrenal quarters incubated under basal conditions and with serotonin (5-HT), ACTH, or both agents added to the medium. Mean values of two flasks ± range. (For methodology see Müller, 1965a; Müller and Weick, 1967)

synthesis. This is consistent with the observation (Fig. 11, Table 2) that it had no cumulative effect when added to the incubation medium together with ACTH or potassium ions in doses which by themselves lead to a maximal stimulation of aldosterone biosynthesis; ACTH and potassium ions were shown to act also on the conversion of cholesterol to pregnenolone in the aldosterone biosynthetic pathway. Serotonin did not stimulate aldosterone production when unlabelled pregnenolone was added in substrate amounts (40 µg/ml) to the incubation medium (Müller, unpubl.). Since serotonin acts on a biosynthetic step which is common to the aldosterone and corticosterone pathway but only stimulates aldosterone production, it has been assumed to act preferentially on zona glomerulosa cells, as has recently been confirmed by incubation of separated zones of the rat adrenal cortex (MÜLLER, 1971).

a) Related Compounds

Testing a series of indole derivatives in equimolar concentrations, ROSENKRANTZ and LAFERTE (1960) found only bufotenine (5-hydroxy-N,N-dimethyltryptamine) to have a steroidogenic activity comparable to that of serotonin. According to JOUAN and SAMPEREZ (1964) tryptamine, 5-hydroxytryptophan, N-methyltryptamine, 5-methoxytryptamine, melatonin (N-acetyl-5-methoxytryptamine and psilocybin (4-phospho-N,N-dimethyltryptamine) have no aldosterone-stimulating activity, which is only partially consistent with our own findings (MÜLLER and ZIEGLER, 1968). We found bufotenine and 5-methoxytryptamine to stimulate aldosterone production by more than 100% above mean control values

Table 6. *Aldosterone-stimulating activity of serotonin and chemically or pharmacologically related compounds. The lowest concentration that stimulated aldosterone production of quartered rat adrenals by 100% above mean control values is added in parentheses. Compounds listed as inactive stimulated aldosterone production by less than 20%.* (From MÜLLER and ZIEGLER, 1968)

Aldosterone-stimulating activity	No aldosterone-stimulating activity at 10^{-5} mol/l	
Serotonin (10^{-7} M)	Melatonin	Histamine
Bufotenine (10^{-6} M)	5-hydroxy-3-indole-	6-methoxytetra-
5-methoxytryptamine (10^{-6} M)	acetic acid	hydroharman
L-5-hydroxytryptophan (10^{-5} M)	L-tryptophan	Adrenaline
	Tryptamine	Noradrenaline
	Tyramine	L-thyroxine

when added in a dose of 10^{-6} mol/l, whereas L-5-hydroxytryptophan had a similar activity in a 10^{-5} molar concentration. At this dose level, melatonin, 5-hydroxyindoleacetic acid, tryptamine and L-tryptophan had no effcet on aldosterone production. Tyramine, histamine, adrenaline, noradrenaline and L-thyroxine were also inactive (Table 6).

b) Serotonin Antagonists

Methysergide and BC-105 (a serotonin antagonist structurally related to cyproheptadine, Sandoz Ltd. Basle) both in a concentration of 16.7 µg/ml decreased base-line aldosterone production by approximately 20% and completely inhibited the effect of serotonin

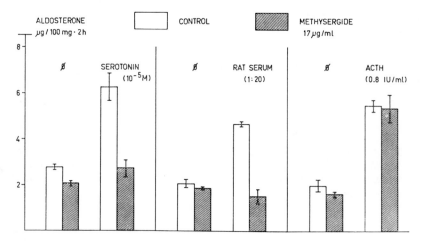

Fig. 12. Effect of methysergide bimaleinate on *in vitro* aldosterone production by rat adrenal quarters and its stimulation by serotonin, rat serum, and ACTH. Mean values of two flasks ± range. (Data from MÜLLER and ZIEGLER, 1968)

$(1.7 \times 10^{-5}\text{M}$, MÜLLER and ZIEGLER, 1968, Fig. 12). The steroidogenic effect of serotonin was not impaired by reserpine (42 µg/ml) or dexamethasone (42 µg/ml) (MÜLLER, 1970).

8. Specific Inhibitors of Steroidogenic Enzymes

In the following, only a few inhibitors of steroidogenic enzymes which interfere with aldosterone biosynthesis will be discussed. For a review of other inhibitors of corticosteroid biosynthesis see KAHNT and NEHER (1966b), and GAUNT et al. (1965, 1968).

a) Metopirone[10]

According to KAHNT and NEHER (1962) Metopirone not only inhibits, 11β-hydroxylation but is even more active in suppressing the 18-hydroxylation and possibly also the 18-hydroxydehydrogenation step in the aldosterone biosynthetic pathway. Aldosterone biosynthesis from endogenous precursors by beef adrenal homogenate was inhibited by a dose of Metopirone which was 50% lower than the one needed to induce a comparable suppression in the production of corticosterone, cortisol and 18-hydroxycorticosterone. Suppression of aldosterone secretion by rats *in vivo* was observed after intravenous injection of 0.1—0.3 mg/kg of Metopirone, whereas the dose needed to inhibit 18-hydroxy-11-deoxycorticosterone and corticosterone secretion was 1 mg/kg and 3 mg/kg, respectively. STACHENKO and GIROUD (1964) found that a low dose of Metopirone which was not sufficient to impair cortisol and corticosterone biosynthesis by beef adrenal zona glomerulosa slices almost completely inhibited aldosterone production from endogenous precursors as well as from added precursor steroids (progesterone, 11β-hydroxyprogesterone, deoxycorticosterone, corticosterone) and blocked the conversion of ^{14}C-labelled corticosterone to aldosterone and 18-hydroxycorticosterone.

The aldosterone-suppressive activity of Metopirone was also demonstrated in studies carried out with normal men as well as in patients with primary or secondary aldosteronism and Cushing's syndrome (COPPAGE et al., 1959; TUCCI et al., 1967). Aldosterone secretion was markedly inhibited in all these subjects, even when

---

10 Metopirone = Metyrapone = Su-4885 = 2-methyl-1,2-bis-(3'pyridyl)-1-propanone.

their sodium intake was restricted. Sodium diuresis and potassium retention was observed when Metopirone was administered together with prednisone or dexamethasone, and an increase in deoxycortico-sterone secretion was prevented.

According to TRAIKOV et al. (1969), *reduced Metopirone* (2-methyl-1, 2-bis(3'-pyridyl)-1-propanol) is an even more selective inhibitor of aldosterone biosynthesis than Metopirone. When added at a low concentration to the medium in which rat adrenal glands were incu-bated, it strikingly inhibited the conversion of added deoxycortico-sterone to aldosterone and 18-hydroxycorticosterone, did not affect the conversion to corticosterone and even increased the conversion to 18-hydroxy-11-deoxycorticosterone.

b) Su 8000, Su 9055, and Su 10603 [11]

In beef adrenal homogenate these three substances inhibited the production of aldosterone, 18-hydroxycorticosterone and cortisol from endogenous precursors but stimulated the production of corticosterone (KAHNT and NEHER, 1962). Whereas Su 8000 and Su 10603 inhibited 17α-hydroxylation most actively, Su 9055 exer-ted a more marked suppressing effect on aldosterone biosynthesis. This substance also blocked the conversion of added pregnenolone and progesterone to aldosterone and the incorporation of ^{14}C-la-belled corticosterone into aldosterone. Addition of Su 8000 to rat adrenal incubates resulted in a 50% decrease in aldosterone produc-tion, a small increase in corticosterone production and a marked increase in deoxycorticosterone production (Fig. 13). In the pre-sence of Su 8000 a high potassium concentration stimulated mainly deoxycorticosterone production instead of aldosterone production.

In normal human subjects on a restricted sodium intake, in patients with cirrhosis of the liver with ascites and in cases of primary aldo-steronism, administration of Su 9055 (4.8 g/d) resulted in a con-siderable suppression of aldosterone secretion and a striking increase in corticosterone secretion, even when ACTH secretion was supp-ressed by dexamethasone, and in natriuresis (BLEDSOE et al., 1964).

---

11 Su-8000 = 3-(6-chloro-3-methyl-2-indenyl)-pyridine; Su-9055 = 3-(1, 2, 3, 4-tetrahydro-1-oxo-2-naphthyl)-pyridine; Su-10603 = 7-chloro-3, 4-dihydro-2(3-pyridyl)-1(2H)-naphthalenone.

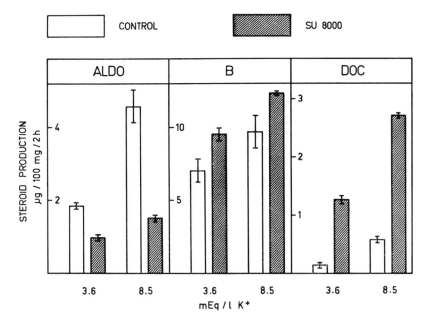

Fig. 13. Effect of Su 8000 (10 μg/ml) on the production of aldosterone (ALDO), corticosterone (B), and deoxycorticosterone (DOC) by rat adrenal quarters incubated in a medium with a low (3.6 mEq/1) and a high (8.5 mEq/1) potassium concentration. Mean values of two flasks ± range. (For methodology see MÜLLER, 1965a; MÜLLER and WEICK, 1967) Su 8000 was a gift from Ciba AG, Basel.

c) Aminoglutethimide

Aminoglutethimide equally inhibited the biosynthesis of cortisol, corticosterone and aldosterone by beef adrenal homogenate (KAHNT and NEHER, 1966b). This effect is due to blocking of the conversion of cholesterol to pregnenolone in adrenocortical tissue according to studies carried out *in vivo* and *in vitro* by DEXTER et al. (1967). Aminoglutethimide directly interferes with adrenal enzymatic cholesterol side-chain fission at least partially by inhibiting 20α-hydroxylation of cholesterol (KAHNT and NEHER, 1966b; COHEN, 1968).

In normal human subjects, administration of aminoglutehimide resulted only in small decreases in cortisol secretion rate and plasma cortisol level (FISHMAN et al., 1967). Consistently observed increases in plasma ACTH levels may have partially abolished the cortisol-suppressing effect of aminoglutethimide; blocking of cortisol secretion was more clearly manifested in a patient with autonomous cortisol secretion by an adrenocortical adenoma. In all normal subjects as well as in patients with Cushing's syndrome or primary aldosteronism, aminoglutethimide induced a striking fall in aldosterone secretion, which was accompanied by sodium diuresis and potassium retention.

9. Inhibitors of Protein Synthesis

a) Actinomycin D

Actinomycin D (10 µg/ml), added to the medium in which quartered rat adrenals were incubated, inhibited both the incorporation of ^{14}C-leucine into adrenal protein and the production of aldosterone by 38% (BURROW et al., 1966). At an equal concentration, actinomycin D inhibited the basal production of aldosterone by rat adrenal tissue by an average of 37% and that of corticosterone by 10% in studies carried out in our laboratory (Table 7). Actinomycin D did

Table 7. *Effect of actinomycin-D (10 µg/ml) on aldosterone and corticosterone production by rat adrenal quarters in vitro. Actinomycin D was added to the preincubation medium and the incubation medium. Mean values of two flasks ± range.*
(For methodology see MÜLLER, 1965a)

| Addition to incubation medium[a] | Steroid production (µg/100 mg/2 h) | | | |
| | Aldosterone | | Corticosterone | |
	Control	Actino-mycin D	Control	Actino-mycin D
—	1.61±0.05	0.96±0.04	7.25±0.39	6.26±0.19
KCl[b]	3.84±0.13	2.94±0.22	10.86±0.23	9.43±0.20
—	1.72±0.19	1.13±0.14	8.89±0.45	8.07±0.39
ACTH[c]	3.74±0.35	3.91±0.60	57.68±0.78	53.95±1.86

[a] 3.6 mEq K+/l. [b] 8.5 mEq K+/l. [c] 5 IU/flask.

not impair the stimulation of aldosterone production by potassium ions or the stimulation of the production of both steroid fractions by ACTH. It has been previously shown by FERGUSON and MORITA (1964) and by HALKERSTON et al. (1965) that actinomycin D does not block the steroidogenic response to ACTH of rat adrenals *in vitro*, whereas it completely blocks the stimulatory effect of ACTH on corticosteroid production by cow adrenal slices (FARESE, 1966).

b) Cycloheximide

Addition of cycloheximide (10^{-3} M) to rat adrenal incubates resulted in a 84% inhibition of amino acid incorporation into adrenal protein and a 63% decrease in aldosterone production (BURROW et al., 1966). We found that cycloheximide induced a 40—50% decrease in the basal production of aldosterone, corticosterone and deoxy-corticosterone by rat adrenal quarters and completely inhibited the steroidogenic response to serotonin and potassium ions (Table 8). On the other hand, when pregnenolone was added in substrate amounts to the incubation medium, similar increments in the output of all three corticosteroid fractions were observed in the absence and in the presence of cycloheximide. These findings indicate that cycloheximide blocks aldosterone production at an early step of the biosynthetic pathway, i. e. a stage preceding the formation of pregnenolone, and does not affect 11β-hydroxylation or the conversion of corticosterone to aldosterone. Administration of cycloheximide to rats *in vivo* blocked the increase in corticosterone production in response to ACTH stimulation by preventing the conversion of cholesterol to pregnenolone in the adrenal cortex (DAVIS and GARREN, 1968).

c) Puromycin

Puromycin (7.5×10^{-4} M) significantly reduced the *in vitro* production of aldosterone, corticosterone and deoxycorticosterone, but not of 18-hydroxy-11-deoxycorticosterone, by rat adrenal tissue (KITTINGER 1964). Puromycin completely inhibited the stimulation of corticosterone, 18-hydroxy-11-deoxycorticosterone and deoxy-

Table 8. *Effect of cycloheximide (10^{-3} M) on corticosteroid production by rat adrenal quarters in vitro. Cycloheximide was added only to the medium used for the final incubation. Mean values of two flasks ±range.* (For methodology see MÜLLER, 1965a; MÜLLER and WEICK, 1967)

Addition to incubation medium[a]	Steroid production (μg/100 mg/2 h)					
	Aldosterone		Corticosterone		Deoxycorticosterone	
	Control	Cycloheximide	Control	Cycloheximide	Control	Cycloheximide
—	2.19±0.09	1.29±0.07	8.87±0.03	4.65±0.38	0.37±0.03	0.25±0.01
Serotonin[b]	5.12±0.18	1.39±0.06	8.73±0.31	5.07±0.43	0.48±0.03	0.26±0.01
—	2.76±0.02	1.48±0.07	7.17±0.50	4.02±0.10		
KCl[c]	5.36±0.37	1.62±0.04	7.18±0.06	3.72±0.52		
—	4.14±0.22	2.28±0.30	5.02±0.29	3.12±0.17	0.39±0.20	0.18±0.02
Pregnenolone[d]	8.28±1.31	5.78±0.19	12.78±2.50	10.13±1.83	3.47±0.13	3.31±0.78

a 3.6 mEq K⁺/l.   b 5-hydroxytryptamine creatinine sulfate, 1.7×10^{-5} M.   c 8.5 mEq K⁺/l.   d 300 μg/flask.

corticosterone production by ACTH; aldosterone production was not stimulated by ACTH in these experiments in the absence of puromycin either. According to Burrow et al. (1966), puromycin (10^{-3} M) inhibited aldosterone biosynthesis in rat adrenal tissue from endogenous precursors by 88% and corticosterone production by 26%. It also blocked the conversion of added progesterone to aldosterone and 18-hydroxycorticosterone, whereas it slightly increased the conversion of progesterone to corticosterone. This finding indicated that puromycin impaired particularly the 18-hydroxylation step of aldosterone biosynthesis. However, addition of large amounts of corticosterone (200 µg/ml) to the incubation medium prevented the inhibition of aldosterone production by puromycin. Moreover, puromycin did not impair a marked increase in aldosterone production in response to added NADP and glucose-6-phosphate.

10. Steroid Hormones

a) Corticosteroids

The conversion of corticosterone to aldosterone and 18-hydroxycorticosterone by sheep adrenal mitochondria was inhibited by small amounts of *18-hydroxycorticosterone* but not by *aldosterone* or *18-hydroxy-11-deoxycorticosterone* (Raman et al., 1966). A large number of different natural and synthetic C_{21}-steroids was found to decrease by 50% or more the conversion of tritiated cholesterol to aldosterone by beef adrenal homogenates when added in high concentrations (150 µg/ml). These steroids had varying effects on the biosynthesis of cortisol and corticosterone (Kahnt and Neher, 1966a). *Cortisol* and *cortisone* were the most active corticosteroids in suppressing aldosterone biosynthesis; they were effective at a concentration of 30 µg/ml. Cortisol (200 µg/ml) inhibited the production of aldosterone in rat adrenal quarters by 45% (Burrow et al., 1966). These investigators proposed that inhibition of aldosterone production by cortisol could be due to inhibition of protein synthesis. According to studies with rat adrenal mitochondria, cortisol blocked the conversion of corticosterone to aldosterone and 18-hydroxycorticosterone (Burrow, 1968). *Dexamethasone* (42 µg/ml) did not directly influence the production of aldosterone and corticosterone by rat adrenal quarters and did not inhibit stimulation of aldosterone biosynthesis by serotonin (MÜLLER, 1970).

b) Androgens and Estrogens

Small amounts of natural androgens and estrogens did not influence the conversion of corticosterone to aldosterone and 18-hydroxycorticosterone by sheep adrenal mitochondria (RAMAN et al., 1966). The biosynthesis of aldosterone from tritiated cholesterol by beef adrenal homogenate was blocked by several natural and synthetic estrogens and androgens (KAHNT and NEHER, 1966a). The most active inhibitors among these compounds were *19-nortestosterone* and *19-hydroxytestosterone*, which decreased aldosterone production at a concentration of 2 μg/ml and 6 μg/ml, respectively. A low dose of *17α-methyltestosterone* (5 μg/ml) selectively inhibited the conversion of labelled progesterone and deoxycorticosterone to aldosterone and 18-hydroxycorticosterone by rat adrenal quarters (REMBIESA et al., 1967). At a higher concentration (20 μg/ml) this steroid also inhibited the incorporation of ^{14}C-progesterone into corticosterone and 18-hydroxy-11-deoxycorticosterone. However, it did not significantly decrease the conversion of corticosterone to aldosterone at either dose level. *Testosterone* (200 μg/ml) inhibited the conversion of labelled corticosterone to aldosterone and 18-hydroxycorticosterone by rat adrenal mitochondria (BURROW, 1968). When dog adrenal glands were perfused *in vivo* with *dehydroepiandrosterone* (30 μg/ml), the conversion of radioactive deoxycorticosterone and 11-deoxycortisol to corticosterone and cortisol, respectively, were markedly decreased, whereas the conversion of deoxycorticosterone to aldosterone was not affected (FRAGACHAN et al., 1969). However, slices of adrenals which had been perfused with dehydroepiandrosterone *in vivo* produced less aldosterone *in vitro* than slices of control adrenals.

11. Ouabain

Ouabain $(5.5 \times 10^{-5}$ M$)$, added to the medium in which cortex slices of dog adrenals were incubated, significantly inhibited aldosterone production (CUSHMAN, 1969). It also prevented the stimulation of aldosterone production which was induced when the potassium concentration in the medium was raised from 5.8 to 9.8 mEq/l. Under the experimental conditions used, addition of ouabain resulted in decreases in tissue uptake of ^{42}K, in tissue potassium content, and in potassium exchange between tissue and medium. WELLEN and BENRAAD (1969a) found that ouabain in low concentrations (10^{-7} M) markedly inhibited the *in vitro* production of corticosterone by outer slices of calf adrenals but not by a mitochondrial preparation of the same tissue. Simultaneously, it blocked active cation transport across cell membranes resulting in a loss of intracellular potassium and a gain in intracellular sodium. Ouabain was found to be less active on rat adrenal tissue (WELLEN and BENRAAD, 1969b). At a concentration of 10^{-5} moles/l it had no effect on corticosterone production by rat adrenal quarters; at a higher concentration of 10^{-4} moles/l it decreased corticosterone biosynthesis by 13%. We observed that a very high dose of ouabain (5×10^{-4} M) resulted in some striking effects on aldosterone biosynthesis by quartered rat adrenals (Table 9). Thus, under basal conditions of incubation (3.6 mEq K+/l) ouabain slightly stimulated aldosterone production by an average of 40%. However, it completely blocked increases in aldosterone production in response to a high potassium concentration in the medium or to added serotonin, and it partially blocked the aldosterone-stimulating effect of ACTH. On the other hand, ouabain did not markedly affect corticosterone production.

Table 9. *Effect of ouabain (5×10^{-4} M) on corticosteroid biosynthesis by rat adrenal quarters in vitro. Ouabain was added only to the medium used for final incubations. Mean values of two flasks ± range.* (For methodology see MÜLLER, 1965a)

| Addition to incubation medium[a] | Steroid production (μg/100 mg/2 h) | | | |
| | Aldosterone | | Corticosterone | |
	Control	Ouabain	Control	Ouabain
—	1.24 ± 0.12	1.47 ± 0.07	8.55 ± 1.15	8.27 ± 0.25
Serotonin[b]	3.33 ± 0.44	1.48 ± 0.01	10.06 ± 0.40	8.80 ± 0.07
—	1.13 ± 0.01	1.91 ± 0.02	5.92 ± 0.12	7.92 ± 0.37
KCl[c]	4.04 ± 0.49	1.95 ± 0.12	6.19 ± 0.78	8.64 ± 0.46
—	1.46 ± 0.04	2.06 ± 0.29	6.54 ± 0.16	7.33 ± 0.21
ACTH[d]	6.69 ± 0.86	3.30 ± 0.34	69.52 ± 0.22	54.65 ± 1.15

[a] 3.6 mEq K+/l.   [b] 5-hydroxytryptamine creatinine sulfate, 1.7×10^{-5} M.
[c] 8.1 mEq K+/l.   [d] 5 IU/flask.

V. Alterations in Aldosterone Biosynthesis and Secretion in Long-Term Experiments and Diseases

1. Alterations in Sodium Balance

Restriction of sodium intake or loss of body sodium by diuresis, sweating, peritoneal dialysis or sequestration of saliva are very potent stimuli of aldosterone secretion. In reverse, increased sodium intake or parenteral administration of sodium result in diminished aldosterone production. This close correlation between sodium balance and aldosterone secretion was first observed in man (LUETSCHER and AXELRAD, 1954) and has since been found in a great number of different animal species including the toad (CRABBE, 1963) the bullfrog (ULICK and FEINHOLTZ, 1968), the rat (SINGER and STACK-DUNNE, 1955), the rabbit (BLAIR-WEST et al., 1968), the dog (ROSNAGLE and FARRELL, 1956), the sheep (BLAIR-WEST et al., 1963), the opossum (JOHNSTON et al., 1965b) the wombat and the kangaroo (COGHLAN and SCOGGINS, 1967). Since sodium retention is the primary biological effect of aldosterone, the physiological importance of the adaptation of aldosterone secretion to alterations in sodium balance is obvious. However, the physiological and biochemical mechanisms involved in this regulation are only partially known in spite of extensive studies carried out in this field by a great number of investigators. The present state of ignorance and controversy is most likely due to the great complexity which appears to pertain to all parts of a possible control system: stimuli, receptors, afferent signals, control organs, efferent signals and response of the adrenal cortex. The physiological strains induced by sodium depletion are numerous including at various stages decreased extracellular fluid volume, decreased total and circulating plasma volume, decreased arterial blood pressure, decreased arterial blood flow, potassium retention and alterations of electrolyte concentrations in the extra- and intracellular fluids. It is difficult to evaluate the relative importance of all these different parameters of sodium loss in eliciting increases in aldosterone production. Perhaps, the diversity of physiological effects of sodium deficiency may explain the diversity of morphological and functional changes it induces in the adrenal cortex, which will be briefly reviewed in the following paragraphs.

a) Morphological Changes

In the rat and—less predictably—in man sodium deficiency results in hypertrophy of the zona glomerulosa (DEANE et al.,1948; PESCHEL and RACE, 1954). A significantly increased zona glomerulosa width was found in rats which had been kept on a sodium-deficient diet for only two days (HARTROFT and EISENSTEIN, 1957). Zona glomerulosa thickness was found to be increased by approximately 100% within one to two weeks of sodium deficiency (DEANE et al., 1948; HARTROFT and EISENSTEIN, 1957; MARX and DEANE, 1963; MÜLLER and HUBER, 1969). Hypertrophy increases successively for several weeks. However, zona glomerulosa of rat adrenals was found to be wider after three weeks of sodium deprivation than after 48 weeks (COHEN, 1965). Results on the width of the zona fasciculata and adrenal weight during sodium deficiency are conflicting. Whereas DEANE et al. (1948) found zona fasciculata width to be normal, HARTROFT and EISENSTEIN (1957) and MARX and DEANE (1963) found it to be decreased to the same extent as zona glomerulosa was increased. In severe sodium deficiency decreases in the width of the inner zones and in total adrenal weight may reflect overall growth retardation (EISENSTEIN and HARTROFT, 1957).

In the rat, the increase of zona glomerulosa width induced by sodium deficiency is independent of the pituitary gland. In chronically hypophysectomized animals the zona glomerulosa was broader than in intact rats and its width was further increased during sodium deficiency (DEANE et al., 1948; PALMORE and MULROW, 1967). Interestingly, in these animals aldosterone secretion *in vivo* (SINGER and STACK-DUNNE, 1955; PALMORE and MULROW, 1967) and aldosterone production by adrenal tissue *in vitro* (LEE and de WIED, 1968) was much lower than in intact sodium-deficient rats.

b) Histochemical Changes

Sodium deficiency results in alterations of the lipid content of zona glomerulosa cells, which are time-dependent. ELEMA et al. (1968a) depleted rats of sodium by peritoneal dialysis with 5% glucose solution. After 3 hours there was a marked lipid depletion in all zones of the adrenal cortex; after 27 hours lipid depletion continued in the zona glomerulosa whereas repletion occured in the two inner zones. Depending on severity or duration of sodium deficiency, lipid droplets in the zona glomerulosa cells become smaller or disappear completely (DEANE et al., 1948). On the other hand, COHEN (1965) found the zona glomerulosa cells to be filled with large lipid droplets in rats which had been kept on a sodium-deficient diet for 48 weeks. It appears to be very difficult to correlate lipid content with steroidogenic activity, since decreased as well as increased secretory activity of the zona glomerulosa may both be associated with lipid depletion (DEANE et al., 1948).

Histochemical changes which accompany increases in zona glomerulosa width during sodium deficiency include increased activity of Δ^5-3β-hydroxysteroid dehydrogenase, glucose-6-phosphate dehydrogenase, isocitrate dehydrogenase and NADPH tetrazolium reductase (MARX and DEANE, 1963; COHEN, 1965; ELEMA et al., 1968a). These changes in enzyme activity could be observed 27 (but not 3) hours after induction of sodium depletion by peritoneal dialysis (ELEMA et al., 1968a) and were prevented by a preceding bilateral nephrectomy (ELEMA et al., 1968b).

c) Altered Sensitivity to Aldosterone-Stimulating Substances

A number of investigators have found that the response of the adrenal cortex in aldosterone output to stimulation by angiotensin II, ACTH, potassium ions and serotonin, respectively, was dependent on sodium intake (THORN et al., 1955; MULLER et al., 1956; VENNING et al., 1962; BLAIR-WEST et al., 1963; MÜLLER, 1965a; 1968; CANNON et al., 1966; CSANKY et al., 1968; KINSON and SINGER, 1968; MÜLLER and HUBER, 1969). Generally it was observed that the response of the adrenal cortex in aldosterone production to an identical dose of one of these aldosterone-stimulating agents was increased during sodium deficiency or decreased during sodium loading. However, according to KINSON and SINGER (1968) only the response to angiotensin II but not the response to ACTH was enhanced by sodium deficiency in rats. Sodium deficiency markedly decreased the response of the adrenal cortex to exogenous angiotensin II in sheep (BLAIR-WEST et al., 1969a). A 48-hour infusion of angiotensin II induced only a transient increase in aldosterone secretion in hypophysectomized dogs which were kept on a high sodium diet (200 mEq/d) in spite of a sustained increase in blood pressure (DAVIS et al., 1969). A similar dose of angiotensin led to a sustained elevation of aldosterone, corticosterone and cortisol secretion in animals on a normal sodium intake (60 mEq/d).

Detailed studies on the effect of sodium deficiency on the sensitivity of the adrenal cortex of the dog to varying doses of stimulating agents have been carried out by the group of Ganong (GANONG et al., 1965, 1966; GANONG and BORYCZKA, 1967). These investigators demonstrated that a comparable increment in aldosterone secretion was elicited by a lower dose of angiotensin or ACTH in dogs kept on a sodium-deficient diet than in sodium-replete dogs. The increased sensitivity became manifest after 5 days of sodium restriction; at that time, hypertrophy of the zona glomerulosa was not yet detectable (GANONG et al., 1967a). Increased responsiveness in aldosterone secretion was not associated with altered responsiveness in 17-hydroxycorticosteroid output to ACTH or angiotensin. The alterations in sensitivity of the adrenal cortex observed in sodium deficiency could be reproduced in sodium-replete dogs by the administration of homologous renin for five days (GANONG et al., 1967a).

d) Sites of Action in the Biosynthetic Pathway

In intact dogs, sodium depletion resulted in a selective increase of aldosterone secretion without any alteration in glucocorticosteroid output; in hypophysectomized dogs, sodium deficiency induced also a significant increase in corticosterone and cortisol production (BINNION et al., 1965; W. W. DAVIS et al., 1968). This indicated that sodium deficiency

1. stimulated steroid production in the zona fasciculata as well as in the zona glomerulosa

2. and acted at an early stage of steroid biosynthesis, common to the production of aldosterone and the two glucocorticosteroids, i.e. at a biosynthetic step preceding the formation of progesterone.

Studies carried out in men, sheep and rats have yielded conflicting evidence on the influence of sodium balance on steroidogenesis in the zona fasciculata, but have largely confirmed that sodium deficiency stimulates an "early" step in the aldosterone biosynthetic pathway. In human subjects, whose endogenous ACTH secretion was suppressed by a constant dose of dexamethasone, sodium depletion led to parallel increases in aldosterone and corticosterone production without a noticeable effect on cortisol secretion rate (BLEDSOE et al., 1966). During treatment with dexamethasone and Metopirone, sodium depletion elicited consistent increases in deoxycorticosterone production. Indirect evidence for a site of action of sodium balance in the initial stages of aldosterone biosynthesis in man can be derived from the data obtained in patients with sodium loss due to an isolated inborn error of aldosterone biosynthesis. In the three cases of 18-hydroxylation deficiency described by VISSER and COST (1964) and by DEGENHART et al. (1966) increased amounts of urinary metabolites of corticosterone were excreted during severe sodium deficiency. Two patients also excreted pathological amounts of deoxycorticosterone in the urine. In the case of 18-hydroxy-dehydrogenation deficiency described by ULICK et al. (1964b) the secretion rate of corticosterone was pathologically elevated during restriction of sodium intake and became normal upon treatment with deoxycorticosterone acetate.

Continuous superfusion and perfusion studies with rat adrenal tissue *in vitro* and sheep adrenals *in vivo*, respectively, showed that aldosterone production was affected by changes in sodium balance at two different sites of the biosynthetic pathway, one before and one after the formation of corticosterone (BANIUKIEWICZ et al., 1968; BLAIR-WEST et al., 1970b). Capsular glands (zona glomerulosa) taken from hypophysectomized rats which had been drinking sodium chloride solution for two days produced less aldosterone, 18-hydroxycorticosterone, and corticosterone during the first 30 minutes of *in vitro* superfusion than corresponding tissue from control animals which had been drinking water. During the later periods of superfusion, tissue from sodium-loaded rats produced less aldosterone and 18-hydroxycorticosterone but more corticosterone and 18-hydroxy-11-deoxycorticosterone than tissue from control rats. In sheep adrenal autotransplants the conversion of infused tritiated corticosterone to aldosterone rose proportionally with the increasing production of aldosterone from endogenous precursors during the onset of mild sodium depletion until the aldosterone secretion rate had reached a value of 10 to 12 μg/h. In more severely sodium-depleted sheep the aldosterone secretion rate rose to higher values, but the conversion of tritiated corticosterone to aldosterone decreased again.

In accordance with the findings of BANIUKIEWICZ et al. (1968), several investigators studying the influence of sodium balance on aldosterone biosynthesis by incubated adrenal tissue (the main incubation generally carried out after a preincubation period of 30 or 60 minutes) observed that sodium deficiency resulted in an increased activity of the conversion of endogenous or added corticosterone to aldosterone. Adrenal cortical slices of sodium-depleted dogs produced significantly more aldosterone and less corticosterone than adrenal slices of sodium-replete dogs; cortisol production was equal in both tissues (W. W. DAVIS et al., 1968). Adrenal tissue of sodium-deficient dogs converted also more unlabelled progesterone and corticosterone (added in substrate amounts to the incubation medium) to aldosterone and incorporated more tritiated progesterone and corticosterone into aldosterone. Adrenals of rats which had been drinking sodium chloride solution produced less aldosterone and more corticosterone when incubated in a medium with a low (3.6 mEq/l) or a high (8.5 mEq/l) potassium concentration, converted less added unlabelled progesterone, 11β-hydroxyprogesterone, deoxycorticosterone and corticosterone to aldosterone and converted less tritiated pregnenolone, progesterone, deoxycorticosterone and corticosterone to aldosterone than adrenals of control animals (MÜLLER, 1968). Adrenals of sodium-deficient rats produced more aldosterone (under basal conditions of incubation and when stimulated by serotonin, potassium ions or ACTH) and less corticosterone (when incubated with serotonin or ACTH) and converted more added progesterone, deoxycorticosterone and corticosterone to aldosterone than adrenals of sodium-replete rats (MÜLLER and HUBER, 1969, Fig. 10, 14). MARUSIC and MULROW (1967b) have shown that adrenal mitochondria of rats kept on a sodium-deficient diet have a greater capacity for converting corticosterone to aldosterone and 18-hydroxycorticosterone than mitochondria of sodium-replete rats. The increased activity of the final steps of aldosterone biosynthesis became manifest after one day and reached a maximum after two days on a low-sodium diet. Provided that there are actually

two enzymes involved in the conversion of corticosterone to aldosterone and that 18-hydroxycorticosterone is the immediate precursor of aldosterone, available evidence on the relative effect of sodium
balance on 18-hydroxylation and 18-hydroxydehydrogenation is
conflicting. BANIUKIEWICZ et al. (1968) found that decreases in

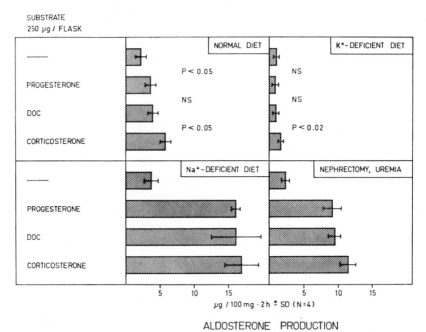

ALDOSTERONE   PRODUCTION

Fig. 14. Aldosterone production by adrenal tissue of various groups of
rats incubated with substrate amounts of unlabelled precursor steroids.
Mean values of two experiments (N = 4) ± standard deviation. P values
of significance were calculated by t tests and refer to differences between
adjacent bars. (NS = not significant at the P = 0.1 level. Data from
MÜLLER and HUBER, 1969)

18-hydroxycorticosterone production by superfused rat capsular
adrenals due to drinking of saline paralleled decreases in aldosterone
production. In man, also, the fractional increases in the secretion
rates of aldosterone and 18-hydroxycorticosterone due to sodium
deficiency were equal (ULICK et al., 1964a; TOUITOU et al., 1970).
But according to *in vitro* studies by VECSEI et al. (1966b), the conversion of tritiated progesterone to aldosterone in adrenal tissue of so-

dium-loaded rats was more markedly decreased than the conversion to 18-hydroxycorticosterone; sodium deficiency induced only an increased incorporation of tritiated progesterone into aldosterone and did not influence the incorporation into 18-hydroxycortico-sterone. Increased sodium intake resulted in a markedly decreased aldosterone production by rat adrenal tissue, but did not consistently influence 18-hydroxycorticosterone production (SHEPPARD et al., 1964).

According to MARUSIC and MULROW (1967b) the conversion of deoxycorticosterone to corticosterone by capsular adrenal mito-chondria was not enhanced by 4 days of sodium restriction. Howe-ver, in studies in which sodium intake of rats was altered for periods of two weeks, we found indirect evidence that sodium balance does not only affect the conversion of corticosterone to aldosterone but induces similar changes in the activity of the 11β-hydroxylation step in the aldosterone biosynthetic pathway (MÜLLER, 1968; MÜL-LER and HUBER, 1969). Adrenals of rats which had been drinking saline produced 100 to 200% more deoxycorticosterone than adrenals of control animals when stimulated *in vitro* by potassium ions or ACTH. In reverse, adrenal tissue of rats which had been kept on a sodium-deficient diet produced significantly decreased amounts of deoxycorticosterone when incubated with or without aldosterone-stimulating substances (see Fig. 10). Moreover, in adrenals of sodi-um-deficient rats aldosterone production was stimulated to the same extent by the addition of substrate amounts of progesterone, deoxycorticosterone and corticosterone to the incubation medium, whereas in adrenals of rats receiving a complete diet aldosterone production was highest when corticosterone was added (Fig. 14). Recent studies have shown that alterations in deoxycorticosterone output due to alterations in sodium balance are limited to the cap-sular portion of the rat adrenal cortex (MÜLLER, 1971, Fig. 2, 5, 8).

e) Role of the Renin-Angiotensin System

Alterations of aldosterone production in response to changes in sodium balance are widely considered to be mediated by the renin-angiotensin system (GROSS et al., 1965; GANONG et al., 1966; DAVIS, 1967). This assumption is mainly based on the following evidence:

1. A low sodium intake or sodium depletion leads to increased peripheral plasma levels of renin in man (BROWN et al., 1964), dog (BINNION et al., 1965), sheep (BLAIR-WEST et al., 1967) and the rat (GROSS et al., 1965). Sodium deficiency also induces an increased renin content of the kidneys (GROSS, 1964) and an elevated level of plasma angiotensin II (SKORNIK and PALADINI, 1964).

2. Angiotensin II has a strong, direct and more or less specific aldosterone-stimulating effect on the adrenal cortex of several animal species and presumably also of man.

3. Bilateral nephrectomy of hypophysectomized sodium-depleted dogs reduced aldosterone secretion by 80 to 90% within one hour (DAVIS et al., 1961, 1966). In sodium-depleted hypophysectomized uninephrectomized sheep, surgical removal of the second kidney did not lead to a reduction of aldosterone secretion within 3 hours (BLAIR-WEST et al., 1964); however, by 10 to 12 hours, aldosterone secretion rate was decreased to the range observed in sodium-replete animals (BLAIR-WEST et al., 1968). At that time aldosterone production rose markedly in response to infusions of angiotensin or renin.

4. Passive transfer of renin antibodies to dogs significantly inhibited aldosterone secretion in sodium-deficient but not in sodium-replete animals (LEE et al., 1965).

Based on this evidence, it can be assumed that the renin-angiotensin system is involved in aldosterone homeostasis during alterations of sodium balance, but the hypothesis stated by BINNION et al. (1965) that "increased activity of the renin-angiotensin system is the primary mechanism leading to hyperaldosteronism during sodium depletion" is in my opinion an oversimplification and debatable in several respects. In experiments designed to test this hypothesis in the sheep, BLAIR-WEST et al. (1969a, b) have made a number of observations which are difficult to reconcile with a primary role of the renin-angiotensin system in inducing and maintaining an increased aldosterone secretion during sodium deficiency. Thus, during rapid correction of a sodium deficit the correlation between plasma levels of aldosterone and of renin and angiotensin was low. Aldosterone secretion increased and decreased normally in response to sodium depletion and repletion in the presence of continuous high levels of peripheral blood angiotensin due to intravenous infusion of synthetic angiotensin. The adrenals of moderately sodium-depleted animals were almost totally unresponsive to angiotensin, and angiotensin infusions did not sustain a high secretion rate of aldosterone. In normal men, an infusion of angiotensin II did not induce the same increase in plasma aldosterone concentration as the one observed after five days of sodium deficiency although the plasma angiotensin II concentration was the same in both instances (BOYD et al., 1969). In addition, the specific alterations in aldosterone biosynthesis seen in sodium-deficient animals can only partially be induced in sodium-repleted animals by administration of exogenous renin and angiotensin or by elevation of endogenous renin levels due to experimental renovascular hypertension. In particular, available experimental evidence does not indicate that angiotensin has a direct stimulating effect on the conversion of corticosterone to aldosterone or on 11β-hydroxylation in the aldosterone biosynthetic pathway.

2. Alterations in Potassium Balance

The influence of potassium balance on aldosterone secretion has not been studied as intensively as the influence of sodium balance, but all available evidence indicates that the total body potassium status may well be as important as the total body sodium status in the overall control of aldosterone production. Increased ingestion or parenteral administration of potassium induces increases, potassium depletion or dietary potassium deprivation induces decreases in aldosterone secretion in man and experimental animals (SINGER and STACK-DUNNE, 1955; FALBRIARD et al., 1955; LARAGH and STOERK, 1957; JOHNSON et al., 1957; GANN et al., 1962, 1964; EILERS and PETERSON, 1964; CANNON et al., 1966).

According to MULLER et al. (1958) potassium balance could influence aldosterone production indirectly by its effects on sodium balance, which lead to changes in the intravascular volume. It is known that potassium loading results in sodium loss (BUNGE, 1873), potassium depletion in sodium retention (BLACK and MILNE, 1952; WOMERSLEY and DARRAGH, 1955). Although potassium balance can in some instances affect aldosterone production through intravascular volume changes and mediation by the renin-angiotensin system, potassium loading stimulated aldosterone also in nephrectomized animals (DAVIS et al., 1963) and in patients with primary aldosteronism (SLATON et al., 1969; CANNON et al., 1966), whose plasma renin is known to be low and to respond poorly to sodium depletion (CONN et al., 1964). Moreover, potassium loading can in certain situations elevate aldosterone secretion rate while decreasing plasma renin activity, as was observed by VEYRAT et al. (1967) when they gave potassium supplements to sodium- and potassium-deficient subjects. This dissociation between plasma renin activity and aldosterone secretion rate was confirmed by DRIESSEN et al. (1969), who studied the effects of potassium loading during a low-sodium diet in normal subjects and patients with essential or reno-vascular hypertension. Potassium balance can influence aldosterone secretion by the direct action of potassium ions on the adrenal cortex, since relatively small changes in plasma potassium concentration—0.5 mEq/l in the sheep (BLAIR-WEST et al., 1962), 1.3 mEq/l in the dog (DAVIS et al., 1963)—can account for marked alterations in aldosterone output (see chapter IV-2-b). Whether potassium balance can influence aldosterone production by additional mechanisms not involving changes in sodium balance or alterations in peripheral plasma potassium concentration is difficult to evaluate. Such possibilities have been proposed by CANNON et al. (1966), who suggested that changes in aldosterone production might rather depend on intracellular adrenocortical potassium content than on plasma potassium concentration, and by GANN et al. (1962), whose experimental data in dogs indicated that potassium could act at some intracranial site to stimulate aldosterone production by a yet unknown mechanism.

Zona glomerulosa width of rat adrenals was found to be decreased by approximately 30% after 2, 5 or 10 weeks of dietary potassium restriction (DEANE et al., 1948; MÜLLER and HUBER, 1969). In reverse, parenteral administration of potassium resulted in a marked increase of zona glomerulosa width within 12 hours (DEANE et al., 1948). Zona fasciculata width appeared to be independent of potassium balance. In this context, it may be of interest that addition of rubidium sulphate or caesium chloride to a sodium- and potassium-deficient diet resulted in zona glomerulosa hypertrophy in rats (BACH et al., 1960).

During short-term incubation of rat adrenal tissue, potassium ions stimulated aldosterone production at an early step of the biosynthetic pathway, preceding the formation of pregnenolone (MÜLLER, 1966), which is consistent with the finding of DAVIS et al. (1963) that adrenal arterial infusion of potassium salts in hypophysectomized, nephrectomized dogs resulted occasionally in small increases of corticosterone secretion. When potassium chloride was added to the drinking fluid of rats for two weeks, the adrenals paradoxically produced less aldosterone under basal conditions of incubation than adrenals of control animals, but equal amounts under *in vitro* stimulation by potassium ions, ACTH or added progesterone, 11β-hydroxyprogesterone, deoxycorticosterone or corticosterone (MÜLLER, 1968). However, in these experiments the rats were kept on a diet with a relatively low sodium (50 mEq/kg) and high potassium (187 mEq/kg) content and the adrenals may have been already maximally stimulated by a positive potassium balance under control conditions. Adrenals of rats kept on a potassium-deficient diet (0.7 mEq/kg K+, 230 mEq/kg Na+) for two weeks produced significantly less aldosterone, but more corticosterone and deoxycorticosterone than adrenals of rats kept on a complete diet (230 mEq/kg K+, 230 mEq/kg Na+) (MÜLLER and HUBER, 1969). In adrenals of potassium-deficient rats, aldosterone production did not respond to stimulation by serotonin, potassium ions or ACTH, whereas the response in deoxycorticosterone production to stimulation by serotonin or potassium ions was increased (Figs. 5, 10). Recent studies (MÜLLER, 1971, Figs. 2, 5, 8) have shown that the increased deoxycorticosterone production induced by potassium deficiency was derived from the zona glomerulosa only and not from the inner zones of the adrenal cortex. Only the addition of corticosterone in substrate amounts resulted in a small but significant increase of aldosterone by adrenals of potassium-deficient rats (Fig. 14). Reduced activity of the enzymes necessary for the final steps of aldosterone biosynthesis, i.e. 11β-hydroxylation and one or both of the reactions involved in the conversion of corticosterone to aldosterone, could account for all these findings According to BOYD et al. (1968), potassium loading for two days resulted in an increased conversion of corticosterone to aldosterone by rat adrenal mitochondria even when sodium losses were compensated. We have also found some preliminary evidence that alterations of the final steps of aldosterone biosynthesis due to changes in potassium balance are not simply mediated by induced changes in sodium balance. Thus, adrenals of rats

kept on a sodium- and potassium-deficient diet for two weeks produced normal amounts of aldosterone but markedly increased amounts
of deoxycorticosterone when incubated with serotonin or in a medium with a high potassium concentration (MÜLLER, unpublished
data). The different site of action on the aldosterone biosynthetic
pathway by potassium ions *in vitro* and alterations of potassium
balance *in vivo* may either be due to the fact that incubations were
carried out for only a few hours whereas alterations in potassium
balance lasted several days or weeks or it may indicate that extra-
adrenal mediators are involved.

3. Exogenous Angiotensin[1] and Renin

If the renin-angiotensin system plays the primary role in the
regulation of aldosterone secretion under physiological conditions
as well as in spontaneous or experimental secondary hyperaldosteronism, as has been assumed by GANONG et al. (1966) and DAVIS
(1967), several criteria should be fulfilled according to AMES et al.
(1965):

1. Angiotensin should stimulate aldosterone secretion in non-
pressor doses or mildly pressor doses, since secondary hyperaldosteronism is generally not accompanied by arterial hypertension.

2. Angiotensin should stimulate aldosterone secretion selectively, since hyperaldosteronism is not usually accompanied by
increased cortisol secretion.

3. Aldosterone secretion should remain elevated as long as the
blood angiotensin level is elevated.

a) Aldosterone Stimulation by Pressor and Non-Pressor Doses

In two normal human subjects 8-hour infusions of angiotensin
at a dose which had no effect on blood pressure elicited a marked
increase in urinary aldosterone excretion (GENEST, 1961). In other

---

1 Synthetic val[5]-angiotensin II-asp[1] β-amide was used in all investigations referred to in this chapter.

instances, pressor doses of angiotensin were needed to stimulate aldosterone secretion in normal men or patients with essential hypertension during short-term or prolonged infusion studies, although the increases in arterial blood pressure were often very small (LARAGH et al., 1960; BIRON et al., 1961, 1962; AMES et al., 1965). In a normal man, a graded response in aldosterone secretion was obtained by pressor doses of human renin (BRYAN et al., 1964).

In intact dogs prolonged infusions of angiotensin for 4 to 11 days at doses which had no effect on blood pressure led to sustained increases in urinary aldosterone excretion and to changes in electrolyte balance that were very similar to those observed during prolonged treatment with mineralocorticosteroids, i.e. transient sodium retention, polyuria, and fecal potassium loss (URQUHART et al., 1963). Single doses of renin extracts from dog kidneys or short-term infusions of angiotensin that stimulated aldosterone secretion in hypophysectomized dogs also increased arterial blood pressure (MULROW et al., 1962).

In sheep, systemically infused angiotensin or renin stimulated aldosterone secretion in pressor doses only (BLAIR-WEST et al., 1962, 1967).

In rats, angiotensin or renin had either no effect on aldosterone production (EILERS and PETERSON, 1964) or stimulated aldosterone secretion in pressor doses (CADE and PERENICH, 1965; MARIEB and MULROW, 1965; BOJESEN, 1966; DUFAU and KLIMAN, 1968a,b; KINSON and SINGER, 1968; MASSON and TRAVIS, 1968). Long-term treatment with angiotensin or renin eliciting increased *in vitro* production of aldosterone and elevated secretion of aldosterone, respectively, resulted also in chronically elevated arterial blood pressure (MARX et al., 1963; MASSON and TRAVIS, 1968).

b) Selectivity of Aldosterone Stimulation

The specificity of the aldosterone-stimulating activity of angiotensin in short-term experiments has been discussed in chapter IV-1. In normal men, prolonged infusions of angiotensin stimulated cortisol secretion only occasionally and only at very high doses (AMES et al., 1965). In patients with cirrhosis of the liver and ascites, very high doses of angiotensin did not stimulate aldosterone secretion but stimulated the secretion rate of cortisol. This effect might have been due to ACTH secretion since it was blocked by dexamethasone administration.

In two conscious dogs with an arteriovenous fistula, prolonged angiotensin infusions for three days led to progressive increases in aldosterone secretion, which were accompanied by small increases in corticosterone secretion; however, corticosterone secretion rates were highest on the first day of angiotensin administration and were within the normal range on the third day of the experiment (URQUHART et al., 1964). During angiotensin infusions for three hours in intact anesthetized dogs, increases in aldosterone secretion were accompanied by comparable fractional increases in corticosterone and deoxycorticosterone secretion (J. O. DAVIS et al., 1968). In two nephrectomized, hypophysectomized dogs, a 4.5-hour infusion of angiotensin stimulated aldosterone secretion significantly, had an equivocal effect on corticosterone secretion and did not affect 17-hydroxycorticosteroid output (MULROW et al., 1962).

Chronic administration of angiotensin to hypophysectomized rats for 36 days had no effect on aldosterone secretion, but at the highest dosage (75 µg/d) a small increment in corticosterone secretion was observed (CADE and PERENICH, 1965). Using a higher dose of angiotensin (250 µg/d) and intact rats, DUFAU and KLIMAN (1968a) observed a significant increase in corticosterone but not in aldosterone secretion after one week of treatment. After two weeks of treatment the secretion rates of both steroids were significantly elevated. At that time, the total adrenal aldosterone content was increased 9-fold, the corticosterone content 2.5-fold.

c) Persistence of Aldosterone Stimulation

In normal human subjects on a normal or high sodium intake, elevation of aldosterone secretion was sustained under prolonged angiotensin infusions for periods up to 11 days (AMES et al., 1965). However, whereas the sensitivity to the pressor effect of angiotensin increased with prolonged treatment, there was no concomitant increase in sensitivity to the aldosterone-stimulating effect of the peptide. In normal dogs and dogs with an arteriovenous fistula, long-term infusions of angiotensin stimulated aldosterone excretion and secretion, respectively, for periods of several days (Urquhart et al., 1963, 1964). In sodium-replete sheep, continuous adrenal arterial or systemic infusion of angiotensin stimulated aldosterone secretion only for a few hours, despite a sustained effect of the peptide on blood pressure (Blair-West et al., 1962). Progressively increasing doses of angiotensin were necessary in order to keep aldosterone secretion at a constant elevated rate for five hours. In rats, long-term treatment with high doses of renin or angiotensin stimulated aldosterone secretion to the same extent as short-term treatment (Masson and Travis, 1968; Dufau and Kliman, 1968a).

d) Effect on Sodium Balance

The interpretation of studies on the effect of exogenous renin or angiotensin on aldosterone secretion is sometimes made difficult by the fact that angiotensin can have a direct natriuretic or anti-natriuretic effect on the kidney (for reviews see Gross et al., 1965; Peart, 1965; Brown et al., 1966). Thus, changes in aldosterone secretion could be due to angiotensin-induced alterations in sodium balance as well as to a direct effect of the peptide on the adrenal cortex. However, in the majority of studies cited above, aldosterone secretion was modified by exogenous angiotensin or renin independently of alterations in sodium balance. Thus, short-term infusions of angiotensin stimulated aldosterone excretion to the same extent in normal human subjects, where it induced sodium retention, and in patients with essential hypertension, where it induced sodium diuresis (Biron et al., 1962). In normal men, prolonged angiotensin infusions stimulated aldosterone equally during the initial four days of sodium retention and during the following period of re-established sodium balance (Ames et al., 1965). Marked sodium diuresis induced by high doses of angiotensin in patients with cirrhosis of the liver and ascites did not result in a further increase in aldosterone secretion (Ames et al., 1965). In the dog also, long-term angiotensin treatment stimulated aldosterone secretion during periods of sodium retention, sodium balance and sodium loss (Urquhart et al., 1963, 1964). Treatment of dogs with homologous

renin for five days resulted in an increased responsiveness in aldosterone secretion to stimulation by ACTH and angiotensin, but
was not accompanied by sodium diuresis (GANONG et al., 1967a).
In sodium-replete sheep, replacement of the sodium loss induced by
an angiotensin infusion did not prevent a transient increase in aldosterone secretion (BLAIR-WEST et al., 1962). In the rat, sodium balance has not been controlled in studies with long-term treatment
with exogenous renin and angiotensin. However, in the normal
rat the sodium-diuretic response to exogenous angiotensin wears
off after one hour, according to GROSS et al. (1965). Chronic administration of hog renin induced similar increases in aldosterone
secretion and zona glomerulosa width in rats receiving tap water
and 1% NaCl solution, respectively, as drinking fluid (MASSON and
TRAVIS, 1968).

e) Renin Antibodies

Prolonged treatment of dogs with hog renin induces the formation of circulating antibodies capable of neutralizing the pressor
activity of both hog and dog renin (HELMER 1958; WAKERLIN, 1958).
GANONG et al. (1963) incubated hog and dog renin with plasma
of dogs immunized with hog renin prior to injection into nephrectomized, hypophysectomized dogs and observed that this treatment
not only inhibited the pressor activity of the enzyme but also its
aldosterone-stimulating and 17-hydroxycorticosteroid-stimulating activity. Passive transfer of antirenin-globulin resulted in an acute
decrease of the aldosterone secretion rate of dogs on a sodium-
deficient diet, but not of dogs on a normal sodium intake (LEE et
al., 1965). However, in dogs actively immunized with hog renin,
aldosterone secretion rate was normal on a high-sodium as well
as on a low-sodium intake (GANONG et al., 1965b). The immunized dogs conserved sodium normally and responded with an increased aldosterone secretion, but only with a slight arterial blood
pressure increase when they were hypophysectomized and subjected to a constriction of the aorta above the renal arteries. The zona
glomerulosa of the immunized dogs was significantly narrower than
normal, and the kidneys contained substantially increased amounts
of renin, the magnitude of the renal renin content correlating with
the plasma antirenin titer.

f) Morphological and Histochemical Alterations of the Adrenal Cortex

Prolonged infusions of angiotensin to intact rats for 7 to 12 days did not alter zona glomerulosa width (MARIEB and MULROW, 1965). Chronic subcutaneous injections of high doses of angiotensin (0.4 mg/100 g/d) to rats resulted in a moderate increase of zona glomerulosa width within two weeks and a marked increase within 4 weeks (MARX et al., 1963). In the hypertrophied zona glomerulosa but not in the inner zones of the adrenal cortex there was an enhanced activity of Δ^5-3β-hydroxysteroid dehydrogenase, glucose-6-phosphate dehydrogenase and succinic acid dehydrogenase (MARX et al., 1963; LAMBERG et al., 1964). A significant increase in zona glomerulosa width was observed in rats that had been treated with hog renin for 12 days (MASSON and TRAVIS, 1968).

4. Renal Hypertension

a) Renovascular and Malignant Hypertension in Man

Hyperaldosteronism secondary to renovascular hypertension was first described by WRONG (1957). Subsequently DOLLERY et al. (1959) observed two patients with malignant hypertension due to unilateral renal artery stenosis with strikingly decreased serum potassium levels, who became normokalemic upon removal of the diseased kidney. Two similar cases were described by LAIDLAW et al. (1960); one had a normal, one an elevated urinary aldosterone level. These investigators also found an increased aldosterone excretion in two out of 5 patients with renovascular hypertension with normal serum potassium levels. Variable increases in aldosterone secretion or excretion rate have since been reported in a large number of patients with hypertension caused by unilateral renal artery stenosis (GOWENLOCK and WRONG, 1962; COPE et al., 1962; SAMBHI et al., 1963; WRONG, 1964; GENEST et al., 1964; CONN et al., 1964; BARRACLOUGH et al., 1965; LARAGH et al., 1966; KAUFMANN et al., 1967; WEIDMANN and SIEGENTHALER, 1967; VEYRAT et al., 1968; GEORGE et al., 1968; LOMMER et al., 1968; BIGLIERI et al., 1968). The overall incidence of hyperaldosteronism in all patients with renovascular hypertension is difficult to evaluate, but may be around 50%. In individual reports it varies from 25% (WEIDMANN and SIEGENTHALER, 1967) to 90% (KAUFMANN et al., 1967). Increased concentration of peripheral plasma renin was found with a similar frequency among patients with renal artery stenosis (BROWN et al., 1964). Simultaneous measurements of peripheral plasma renin concentration or renin activity and aldosterone secretion rate have been carried out in only a limited number of patients. According to LARAGH et al. (1966), values of plasma renin activity were generally proportional to normal or increased aldosterone secretion rates. A good correlation between increases in aldosterone secretion and increases in plasma renin level was also found by BARRACLOUGH et al. (1965) and VEYRAT et al. (1968). On the other hand, only two out of 8 patients of WEIDMANN and SIEGENTHALER (1967) with a pathologically increased peripheral plasma renin activity had also a pathologically increased aldosterone secretion rate. An even higher incidence of hyperaldosteronism was observed among patients with malignant hypertension (27 out of 28) or advanced hypertension (22 out of 28), but not among patients with benign essential hypertension, by LARAGH et al. (1960b, 1966). A similar frequency was found by VEYRAT et al. (1968). According to these two groups of investigators, plasma renin activity was elevated in patients with malignant hypertension at least in proportion to increases in aldosterone secretion and in some instances to a markedly greater extent. The opposite constellation, i.e. elevated aldosterone secretion rate and normal or low plasma renin activity, has as yet not been observed in patients with renovascular or malignant hypertension. This is consistent with the obvious assumption that in these diseases hyperaldosteronism is due to an increased activity of the renin-angiotensin system. However, the causal relation between increased renin activity and increased aldosterone secretion may be more complex and involve alterations in sodium balance. Patients with malignant hypertension frequently have low or decreased serum sodium levels (LARAGH et al., 1960b). When NEWBORG and KEMPNER (1955) treated 159 patients with malignant hypertension with a rice diet, they had to give sodium chloride to 33 patients because of progressive hyponatremia. Acute renal ischemia in a 22-year old man not only led to malignant hypertension with extremely high plasma levels of renin, hyperaldosteronism with hypokalemia and metabolic alkalosis, but also to severe sodium and water depletion and a reduced plasma volume; surgical treatment of the renal artery stenosis corrected all these abnormalities (BARRACLOUGH, 1966). An inverse correlation between plasma renin and sodium concentrations in patients with renal or malignant hypertension has been observed by BROWN et al. (1965). In four out of six patients with hyperaldosteronism due to unilateral renal artery stenosis, serum sodium concentrations were abnormally low, and in all of 5 cases total blood volume, plasma volume and red cell volume were either in the low normal range or decreased (SLATON and BIGLIERI, 1965). Thus, it seems at least theoretically possible that a low serum sodium concentration, a low total body sodium, or both, substantially contribute to increased aldosterone secretion in some patients with renovascular or malignant hypertension. Available evidence does not allow conclusions as to whether the elevated plasma renin level is rather a cause or a consequence of the disturbance of sodium balance.

BIGLIERI et al. (1968) found normal secretion rates of corticosterone and deoxycorticosterone in 2 patients with malignant and 4 patients with renovascular hypertension.

b) Experimental Renovascular Hypertension

Animal studies on the influence of experimental renovascular hypertension on aldosterone secretion have generally confirmed the observations made in human renovascular hypertension cited above. Aldosterone production was found to be independent of blood pressure and was increased only under experimental conditions which also induced a marked increase in the peripheral plasma renin level or in the renin content of the ischaemic kidney.

CARPENTER et al. (1961) induced renal hypertension in dogs by constricting both renal arteries. According to the tightness of the second clamp two different types of hypertension developed. In dogs with a benign form of hypertension, aldosterone secretion remained within normal limits and the renin content of the kidneys was only slightly increased. In dogs with a malignant form of hypertension, aldosterone secretion rose markedly, whereas corticosterone output remained normal; a 10-fold increase of the kidney renin content was observed in this group. Plasma sodium and potassium levels were normal. In dogs in which hypertension was induced by unilateral renal artery constriction and contralateral nephrectomy, aldosterone secretion rate decreased markedly during the phase of increasing arterial blood pressure and remained at subnormal levels after the blood pressure had reached its maximum value (McCAA et al., 1965). In short-term experiments, constriction of one renal artery in intact as well as in unilaterally nephrectomized dogs was promptly followed by a sharp increase in aldosterone secretion without a concomitant change in cortisol or corticosterone output (MULLER et al., 1964).

In sheep, constriction of one renal artery with the opposite kidney left intact resulted in only a temporary increase in peripheral plasma concentration of aldosterone, cortisol and renin (BLAIR-WEST et al., 1968c). After a few days all these parameters returned to normal values, whereas blood pressure remained elevated. Some animals with severe hypertension became sodium-deficient due to renal sodium loss. In these sheep, renin and aldosterone levels did not decrease to normal but rose progressively in proportion to the sodium deficit. When the artery of the remaining kidney in unilaterally nephrectomzied sheep was constricted, hypertension developed, but sodium balance was maintained; plasma renin and aldosterone rose initially but became normal within three days.

SPÄT et al. (1966) constricted the aorta of rats above or below the renal arteries and found that suprarenal constriction resulted in increased, infrarenal constriction in decreased aldosterone and corticosterone production by incubated adrenal tissue. Adrenals of rats with acute renal hypertension, induced by renal capsulation, converted more ^{14}C-labelled progesterone to aldosterone and 18-hydroxycorticosterone than adrenals of control animals (VECSEI et al., 1966b). Adrenals of rats with chronic renal hypertension due to unilateral renal artery constriction showed a normal rate of incorporation of ^{14}C-progesterone into aldosterone but a decreased incorporation into 18-hydroxycorticosterone. However, chronic experimental hypertension after unilateral artery constriction led to a significant 200% increase in aldosterone secretion, but not in corticosterone, secretion according to SINGER et al. (1963) and KINSON et al. (1967). If one renal artery was clamped and the contralateral kidney was removed, aldosterone secretion remained unchanged despite a marked increase in blood pressure. These findings have been confirmed by our own studies (MÜLLER and GROSS, 1969) on aldosterone biosynthesis in adrenal tissue of rats with different forms of experimental renovascular hypertension. Adrenals of rats with one renal artery constricted and the other kidney left intact produced 200% more aldosterone from endogenous precursors and converted 50% more tritium-labelled pregnenolone, progesterone and corticosterone to aldosterone than adrenals of control animals when incubated under basal conditions (Fig. 15). The difference in aldosterone production was less marked when serotonin, KCl or ACTH were added to the incubation medium. The production of corticosterone and deoxycorticosterone (Fig. 16) in adrenals of renal hypertensive rats was almost normal, when incubated with or without aldosterone-stimulating substances[2]. If rats received 1% NaCl solution instead of water as drinking fluid, the rise in aldosterone production was not prevented (Fig. 17). But the increase in aldosterone production seen in the presence of an intact contralateral kidney was partially or completely inhibited, when simultaneously with renal artery constriction the contralateral kidney was removed or the ureter of the ischaemic or the contralateral kidney was ligated. GROSS et al. (1965, 1968) have shown that in experimental renal hypertension a marked rise in the renin content of the ischaemic kidney and in peripheral plasma renin-like activity is dependent on an intact, functioning contralateral kidney. The close correlation observed between increases in aldosterone production and increases in kidney and plasma renin concentration are at least compatible with the assumption that increases in aldosterone production seen in renovascular hypertension are induced by an increased activity of the renin-angiotensin system. Activation of aldosterone biosynthesis by renovascular hypertension appeared to be qualitatively different from activation by sodium deficiency. Whereas sodium deficiency strikingly activated the final

---

2 This contrasted with the markedly decreased production of deoxycorticosterone and the moderately decreased production of corticosterone observed in adrenals of sodium-deficient rats (MÜLLER and HUBER, 1969; Figs. 2, 5, 8, 10) and was rather similar to the pattern of steroid production by adrenals of rats on a sodium- and potassium-deficient diet (MÜLLER, unpublished observations).

stages of aldosterone biosynthesis, i.e. steps between deoxycorticosterone and aldosterone, renovascular hypertension mainly stimulated early steps in the aldosterone biosynthetic pathway prece-

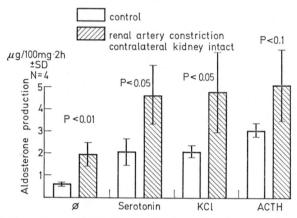

Fig. 15. Aldosterone production *in vitro* by adrenal tissue of normal rats
(white columns) and rats with experimental renal hypertension due to
clamping of one renal artery (shaded columns). Mean values of two
experiments ($N = 4$) $\pm$ standard deviation. *P* values of significance were
calculated by *t* tests. The following substances were added to the incubation
medium: serotonin creatinine sulfate (1.7×10^{-5}M), KCl (final K^+ concentration 8.5 mEq/l), and ACTH (5 IU per flask). (From MÜLLER and
GROSS, 1969)

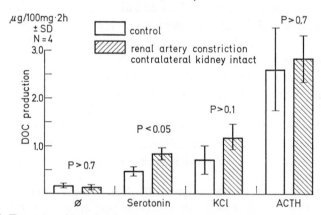

Fig. 16. Deoxycorticosterone production *in vitro* by adrenal tissue of normal rats (white columns) and rats with experimental renal hypertension
due to clamping of one renal artery (shaded columns). (See Fig. 15 for
further explanations. Data from MÜLLER and GROSS, 1969)

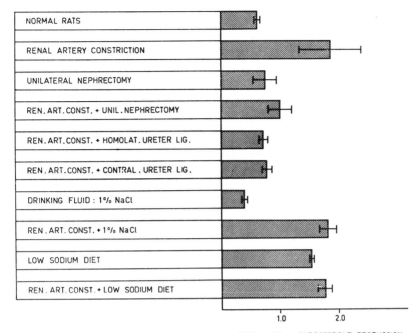

Fig. 17. Aldosterone production by quartered adrenal glands of various groups of rats incubated under basal conditions. Mean values of two flasks ± range. (Data from Müller and Gross, 1969)

ding the formation of pregnenolone. This has been indirectly confirmed by Rapp (1969a) who bred rats selectively for a high or a low juxtaglomerular granularity. Animals with a high juxtaglomerular index had a lower blood pressure but a higher renal and serum renin level; adrenal tissue of these animals produced significantly more aldosterone, corticosterone and deoxycorticosterone than adrenals of rats bred for a low juxtaglomerular index.

5. Bilateral Nephrectomy

Bilateral nephrectomy has been frequently used in experiments carried out in order to evaluate the role of the renin-angiotensin system in the regulation of aldosterone secretion. Generally two different experimental approaches have been used.

In the first type of investigations, the acute effects of nephrectomy on aldosterone secretion were studied in animals with established secondary hyperaldosteronism. Interpretation of these experiments depends greatly on our very limited knowledge of the mode and rate of disposal of renin in the organism. In normal rats endogenous renin disappeared from peripheral blood within one hour after bilateral nephrectomy; in rats with elevated levels of plasma reninlike activity due to unilateral renal artery constriction, the disappearance time was approximately twice as long (SCHAECHTELIN et al., 1964). However, a renin-like enzyme was found in the plasma of four anephric women 1 to 6 months after nephrectomy (CAPELLI et al., 1968). This enzyme may have been of uterine origin. But renin-like activity, acting on homologous substrate at pH 7.5, has also been detected in the plasma of three anephric men (MCKENZIE and MONTGOMERIE, 1969).

In a second type of experiments, animals were first nephrectomized and then subjected to stimuli of aldosterone secretion. In such studies animals often become uremic, and it has to be considered that uremia with hyperkalemia may also be a potent stimulus of aldosterone production.

BALIKIAN et al. (1968) found a high normal plasma concentration and blood production rate of aldosterone in an anephric man; however, these parameters did not change in a normal fashion in response to postural changes.

Nephrectomy resulted in a marked fall of aldosterone secretion in hypophysectomized dogs with hyperaldosteronism due to thoracic inferior vena cava constriction (DAVIS et al., 1961b). In sodium-depleted hypophysectomized dogs removal of the kidneys also induced a striking decrease of the secretion rate of aldosterone, corticosterone and cortisol (DAVIS et al., 1961b, 1966). This was not confirmed by SLATER et al. (1965), who found that in sodium-deficient, hypophysectomized dogs nephrectomy was not followed by a significant change in the secretion rate of aldosterone, corticosterone or cortisol. Hemorrhage markedly increased aldosterone secretion in hypophysectomized dogs but not in hypophysectomized, nephrectomized dogs (GANONG and MULROW, 1961; DAVIS et al., 1961a; SLATER et al., 1965). Bilateral nephrectomy markedly decreased the rise in aldosterone secretion in dogs in response to a lethal dose of *Escherichia coli* endotoxin (WHITE et al., 1967).

Thoracic inferior vena cava constriction significantly stimulated aldosterone secretion in midcollicular-decerebrated, hypophysectomized, pinealectomized sheep, but had no effect in animals which were also bilaterally nephrectomized (BLAIR-WEST et al., 1963). In sodium-replete uninephrectomized, midcollicular-decerebrated sheep removal of the second kidney was followed by a marked decrease in aldosterone production (BLAIR-WEST et al., 1963, 1964). In sodium-deficient sheep, the second nephrectomy did not significantly affect aldosterone secretion, which continued at an elevated rate for several hours. On the other hand, in sodium-deficient, bilaterally nephrectomized sheep, midcollicular decerebration with hypophysectomy and pinealectomy was followed by a substantial fall in aldosterone secretion; however, after all these ablations aldosterone secretion remained elevated above the level found in normal, sodium-replete sheep for periods of 3 to 7 hours. Prolonged studies carried out in conscious sodium-deficient sheep showed that bilateral nephrectomy following hypophysectomy eventually resulted in a reduction of the aldosterone secretion rate to values seen in sodium-replete animals, but in a majority of experiments this occurred 10 to 12 hours after nephrectomy (BLAIR-WEST et al., 1968b). The sustained elevated aldosterone secretion rate was not due to hyperkalemia, which was prevented by appropriate dietary measures. In sodium-replete hypophysectomized sheep aldosterone secretion rate did not alter for several hours after bilateral nephrectomy, until it rose coincidently with a rise in plasma potassium concentration. In these animals, aldosterone secretion rate was stimulated by a renin infusion to values observed in sodium-deficient sheep, but returned to normal within two hours after cessation of the infusion.

In sodium-replete rats acute bilateral nephrectomy was followed by a 30% decrease in aldosterone and corticosterone secretion (EILERS and PETERSON, 1964). When nephrectomy was combined with hypophysectomy, the decrease in aldosterone secretion was approximately 50%. Nephrectomy prevented the stimulation of aldosterone secretion by inferior vena cava constriction. Sodium-deficient hypophysectomized rats secreted aldosterone at a rate twice that of sodium-replete intact animals; bilateral nephrectomy induced only a 25% decrease in aldosterone secretion within two hours. According to MARIEB and MULROW (1965), combined hypophysectomy and bilateral nephrectomy lowered aldosterone secretion

rate by 66% in sodium-replete rats; infusion of angiotensin II restored aldosterone secretion to the preoperative level. Rats that had
been kept on a sodium-deficient diet for 10 days secreted aldosterone
at a markedly elevated rate 6 or 18 hours after bilateral nephrectomy or 2—3 hours after nephrectomy and hypophysectomy (PAL-
MORE et al., 1969). Sustained hypersecretion of aldosterone was
not dependent on a low sodium or a high potassium concentration
in the plasma. If sodium-replete rats were first nephrectomized and
then sodium-depleted by peritoneal dialysis, aldosterone secretion
increased by 100%, corticosterone secretion decreased markedly and
zona glomerulosa width increased slightly. When nephrectomy was
followed by peritoneal sodium loading, aldosterone secretion and
zona glomerulosa width remained normal, but corticosterone secretion decreased to an even greater extent than it did in nephrectomized
sodium-depleted rats. The authors observed no correlation between
plasma sodium concentration and aldosterone secretion of these
animals. Nevertheless, the mean plasma sodium concentration was
considerably higher in the nephrectomized, sodium-loaded group
(165m Eq/l) than in the nephrectomized sodium-depleted group
(147 mEq/l) and plasma potassium was elevated to a mean level
as high as 10 mEq/l in both groups. ELEMA et al. (1968b) showed that
lipid depletion and increases in enzyme activities[3] in the zona glomerulosa induced by peritoneal sodium depletion in intact rats were also
induced by bilateral nephrectomy and were not further affected by
subsequent sodium depletion or sodium loading; serum potassium
concentrations were found to be equally elevated in all groups of
experimental animals. SPÄT et al. (1966a) observed that aldosterone
production by rat adrenals *in vitro* was significantly increased 2 hours
after bilateral nephrectomy. Bilateral nephrectomy prevented a further increase in aldosterone production due to suprarenal constriction of the aorta or a decrease due to infrarenal constriction (SPÄT
et al., 1966b). Infusions of potassium chloride to bilaterally nephrectomized rats resulted in an additional enhancement of aldosterone
production *in vitro* (STURCZ et al., 1967). Adrenals taken from uremic rats 40—48 hours after bilateral nephrectomy produced normal

---

3 isocitrate dehydrogenase, glucose-6-phosphate dehydrogenase, NADP-
tetrazolium reductase, 3β-hydroxysteroid deydrogenase.

amounts of aldosterone and decreased amounts of corticosterone from endogenous precursors and converted ^{14}C-labelled progesterone and tritiated deoxycorticosterone at an increased rate to aldosterone, 18-hydroxycorticosterone and 18-hydroxy-11-deoxycorticosterone, but at a normal rate to corticosterone (STEINACKER et al., 1968). These findings are only partially consistent with our results of incubation studies with adrenal tissue taken from uremic rats 40 hours after nephrectomy (MÜLLER and HUBER, 1969). Adrenal tissue of uremic rats produced significantly increased amounts of aldosterone when incubated in the presence of aldosterone-stimulating substances, such as serotonin, potassium chloride or ACTH, or of precursor steroids (progesterone, deoxycorticosterone, corticosterone) added in substrate amounts to the incubation medium (see Figs. 10, 14). Under stimulation by serotonin or potassium ions, adrenals of uremic rats produced less corticosterone than adrenals of control animals. Base-line deoxycorticosterone production and its stimulation by serotonin and potassium ions were also significantly decreased in adrenals of nephrectomized rats. On the basis of these data we assumed that the final steps in the aldosterone biosynthetic pathway, i.e. 11β-hydroxylation and one or both of the reactions involved in the conversion of corticosterone to aldosterone had been activated either by the nephrectomy or the uremia. We proposed that these changes were most likely due to the marked hyperkalemia, because they were almost exactly the opposite of the alterations of aldosterone biosynthesis induced by potassium deficiency. Chronic renal failure in partially nephrectomized rats with azotemia but normal serum electrolytes did not result in altered aldosterone production by adrenals incubated *in vitro* (HAYSLETT et al., 1969).

Bilateral renal denervation had no effect on aldosterone secretion in sodium-replete rats, but partially inhibited the rise in aldosterone secretion induced by one week of sodium deficiency (KINSON and SINGER, 1969). Two weeks of sodium deficiency stimulated aldosterone secretion to the same extent in rats with denervated kidneys as it did in intact rats. Studies carried out in men after renal homotransplantation have shown that renal denervation does not prevent a marked rise in plasma renin activity and urinary aldosterone excretion in response to dietary sodium restriction (GREENE et al., 1968).

6. Hypophysectomy and Hypopituitarism

In man and experimental animals, such as dogs, sheep, and rats, aldosterone secretion continues in the complete absence of the pituitary gland. A great part of our present knowledge on the physiological mechanisms regulating aldosterone secretion was obtained in experiments with animals which were hypophysectomized in order to abolish the influence of ACTH, a hormone known to stimulate aldosterone production (see chapter IV-4). However, many investigators have found that hypophysectomy or spontaneous hypopituitarism resulted in an impairment of aldosterone secretion under basal conditions or under various physiological stimuli. The extent of this impairment appears to depend greatly on species and the experimental protocol used.

Aldosterone excretion was normal in sodium-replete hypopituitary men, but in some patients it did not respond in a normal fashion to sodium deficiency (LIEBERMAN and LUETSCHER, 1960). In 13 patients with various degrees of hypopituitarism, aldosterone secretion rate was found to be normal and to be stimulated by sodium deficiency to the same extent as in normal subjects (WILLIAMS et al., 1968a). However, these patients showed a greatly diminished response in aldosterone production to the combined stimuli of sodium deficiency and ACTH infusion, whereas patients with an isolated ACTH deficiency due to prolonged treatment with prednisone responded with a normal increment in aldosterone secretion (WILLIAMS et al., 1968b).

Hypophysectomy resulted in an acute decrease of aldosterone secretion by approximately 90% in conscious or anesthetized dogs with hyperaldosteronism due to constriction of the thoracic inferior vena cava (DAVIS et al., 1960a,b). However, in some dogs, aldosterone secretion remained above normal for several days. It was assumed by DAVIS (1961) that the effect of hypophysectomy was due to abolished ACTH secretion, since it was prevented by ACTH infusions. MULROW and GANONG (1961) found that the increment in aldosterone secretion induced by hemorrhage was markedly reduced but not abolished in dogs one hour after hypophysectomy; ACTH infusion stimulated aldosterone secretion to values seen after hemorrhage in intact dogs. In dogs with secondary aldosteronism induced by an infrarenal aortic-caval fistula, hypophysectomy led to a greater fractional decrease in aldosterone secretion than did subsequent bilateral nephrectomy (DAVIS et al., 1964). In hypophysectomized dogs, aldosterone secretion rate was approximately one third that of intact dogs under normal sodium intake as well as under sodium depletion (BINNION et al., 1965). In dogs hypophysectomized 4 weeks previously, neither ACTH nor angiotensin stimulated 17-hydroxycorticosteroid secretion and both agents had a markedly diminished effect on aldosterone production (GANONG et al., 1967). In contrast, dogs with intact pituitaries but treated with large doses of a glucocorticosteroid for 4 weeks showed a normal response in aldosterone output to stimulation by these two agents, although the responses in 17-hydroxycorticosteroid production were also completely absent. It was concluded that a pituitary factor other than ACTH was responsible for the decreased responsiveness in aldosterone secretion in chronically hypophysectomized animals.

In sodium-replete sheep, hypophysectomy combined with pinealectomy and superior collicular decerebration significantly reduced the secretion of cortisol, corticosterone and aldosterone, but in sodium-deficient sheep these operations had in most instances no effect on aldosterone secretion, which continued at its elevated preoperative rate for several hours (BLAIR-WEST et al., 1963). In hypophysectomized sheep that were maintained on cortisone or ACTH treatment for periods of several weeks, sodium deficiency stimulated aldosterone secretion to the same extent as it did in intact animals.

In sodium-replete rats hypophysectomy resulted in a 70% decrease in aldosterone secretion, according to SINGER and STACK-DUNNE (1955). Acute treatment with ACTH restored aldosterone secretion to normal as long as 7 days after hypophysectomy. Maintenance medication with ACTH or growth hormone in hypophysectomized rats had no effect on the response in aldosterone production to acute stimulation with ACTH. Sodium deficiency induced only a small increase in aldosterone production in hypophysectomized rats, which was below the level observed in intact sodium-replete rats. According to EILERS and PETERSON (1964), acute hypophysectomy led to a 30% decrease in aldosterone secretion in sodium-replete rats. In rats that were kept for 2 to 4 weeks on a sodium-deficient diet prior to hypophysectomy, aldosterone secretion rate was twice that of intact animals on normal sodium intake. PALMORE and MULROW (1967) observed that in chronically hypophysectomized rats the response in aldosterone secretion to sodium deficiency was completely abolished; it was partially restored by injections of whole pituitary glands but not by treatment with ACTH, growth hormone, or ACTH and thyroxine. These findings were partially confirmed by LEE et al. (1968) and LEE and de WIED (1968), who studied the influence of hypophysectomy and sodium deficiency on aldosterone production by rat adrenals *in vitro*. They found that hypophysectomy prevented the increase in aldosterone production induced in intact rats by sodium deficiency; adrenal responsiveness was not restored by treatment with ACTH, but by treatment with porcine pituitary powder or by treatment with bovine growth hormone and ACTH. According to newer studies by PALMORE et al. (1970), treatment with a combination of growth hormone and ACTH or with rat pituitary extract completely restored the high level of aldosterone secretion rate in rats that were sodium depleted before hypophysectomy but not in rats that were sodium depleted after hypophysectomy.

Additional experimental and clinical evidence for a possible role of growth hormone in the regulation of aldosterone production is conflicting. Administration of pig, monkey and human growth hormone to hypophysectomized rats *in vivo* led to an increased aldosterone production by adrenal tissue *in vitro* without affecting the biosynthesis of corticosterone (LUCIS and VENNING, 1960; VENNING and LUCIS, 1962). Human growth hormone stimulated the excretion of aldosterone in patients with panhypopituitarism (BECK et al., 1960), but not in normal human subjects (BIGLIERI et al., 1961). FINKELSTEIN et al. (1965) did not observe significant and consistent increases in the aldosterone secretion rate during treatment of hypopituitary dwarfs with human growth hormone. High doses of human growth hormone directly stimulated the output of aldosterone, corticosterone and 17-hydroxycorticosteroids by isolated perfused dog adrenals but this effect may have been due to contamination with ACTH (CUSHMAN et al., 1966).

In the rat, the increased width of the zona glomerulosa due to sodium deficiency is not dependent on the pituitary gland (DEANE et al., 1948; PALMORE and MULROW, 1967). In women with breast cancer, a normal zona glomerulosa was found 9 months after hypophysectomy (JESSIMAN et al., 1959) whereas the two inner zones of the adrenal cortex had completely atrophied.

At present, there is no direct evidence available as to which step of the aldosterone biosynthetic pathway is mainly affected by hypophysectomy. However, the observation that changes in sodium balance influence aldosterone biosynthesis at a stage before as well as at a stage after the formation of corticosterone formation, was made in hypophysectomized rats (BANIUKIEWICZ et al., 1968). Thus neither effect could have depended on an intact pituitary gland.

7. Lesions of the Central Nervous System

It is generally assumed that the central nervous system plays a minor role in the physiological control of aldosterone secretion and that its only important influence on the zona glomerulosa of the adrenal cortex is mediated by pituitary ACTH secretion (MULROW, 1966; DAVIS, 1967). This assumption is based on the finding by DAVIS et al. (1961a) that decapitation of hypophysectomized dogs did not alter aldosterone secretion and did not prevent a marked increase in aldosterone output in response to hemorrhage. Although the presence and integrity of the central nervous system is not essential for aldosterone secretion under basal conditions or in response to various stimuli, there is some evidence indicating that certain areas of the brain can modify aldosterone secretion by unknown mechanisms independent of ACTH secretion.

a) Brain Lesions

According to KRIEGER and KRIEGER (1964) 4 out of 7 patients with *pretectal lesions* did not increase urinary aldosterone excretion in response to sodium restriction. However, none of these subjects exhibited clinical or laboratory evidence of hypoaldosteronism.

In dogs *hypothalamic lesions* had no effect on aldosterone secretion unless they resulted in bilateral injury of the *median evidence*, which was also followed by adrenal atrophy and decreased secretion of 17-hydroxycorticosteroids (DAVIS et al., 1959b; GANONG et al., 1959). *Complete midbrain transection* did not affect aldosterone secretion rate or its stimulation by hemorrhage (DAVIS et al., 1961c). In dogs with secondary hyperaldosteronism due to thoracic inferior vena cava constriction, midbrain transection did not lower the elevated secretion rate of aldosterone if the high venous pressure was maintained.

Midcollicular brain section with removal of the forebrain in hypophysectomized, nephrectomized dogs led to a significant increase in aldosterone, cortisol and corticosterone output (BARBOUR et al., 1965). Whereas the secretion rate of the two glucocorticosteroids remained below the level observed in intact animals, aldosterone was secreted at a rate comparable to that observed in intact or hypophysectomized dogs after hemorrhage, caval constriction or salt depletion.

Discrete lesions in the *dorsal central region of the rostral midbrain* (involving the posterior commissure and the periaqueductal gray substance) markedly decreased aldosterone secretion in cats on a normal or a low sodium intake (TAYLOR and FARRELL, 1962), whereas larger lesions (involving the habenulae, stria medullaris and the pineal gland) obscured this effect at least partially.

In sodium-deficient sheep, *midcollicular decerebration* impaired the rapid decrease of aldosterone secretion in response to a systemic NaCl infusion (BLAIR-WEST et al., 1960, 1963).

In rats, electrocoagulation of the *subcommissural organ* resulted in decreases in aldosterone production *in vitro* as well as in width and nuclear volume of the zona glomerulosa (PALKOVITS et al., 1965). Intraperitoneal injection of saline extracts of the subcommissural organ induced a transient increase of zona glomerulosa nucleus volume (PALKOVITZ and FÖLDVARY, 1963).

b) Pinealectomy

KINSON et al. (1967) found that pinealectomy resulted in a marked increase in aldosterone secretion and a smaller increase in corticosterone production in normal rats. In pinealectomized animals, unilateral renal artery constriction stimulated aldosterone secretion more markedly than it did in intact rats and resulted in an increased output of corticosterone (which was not observed in rats with an intact pineal gland). Sodium deficiency induced the same fractional increase in aldosterone production in intact and in pinealectomized

rats (KINSON and SINGER, 1967). The effect of pinealectomy on aldosterone secretion was a transient one. Three months after the opera-
tion, aldosterone and corticosterone secretion rates had returned to
control levels (KINSON et al., 1968).

Administration of ubiquinone, a constituent of the lipid fraction
of the pineal gland[4], led to a considerable decrease in aldosterone
secretion without a significant effect on cortisol secretion in intact
dogs (FABRE et al., 1965). A hexane extract of beef pineal gland
appeared to inhibit 11β-hydroxylation *in vitro* (LOMMER, 1960).
It decreased the conversion of ^{14}C-labelled progesterone to aldosterone, corticosterone, cortisol and cortisone and enhanced its
incorporation into deoxycorticosterone and 11-deoxycortisol by beef
adrenal slices.

8. Mineralocorticosteroids

Autonomous overproduction of aldosterone by an adrenocortical adenoma, i.e. primary aldosteronism, leads to an impaired secretion of aldosterone by the adjacent and contralateral adrenal
cortex. Marked decreases in aldosterone secretion rate were found
to persist for several weeks after surgical removal of an aldosterone-
producing adenoma (CONN et al., 1964; BIGLIERI et al., 1966a). During this period, aldosterone secretion did not increase in response
to stimulation by ACTH, angiotensin or a sodium-deficient diet.
Sodium restriction resulted in severe urinary sodium loss and hyperkalemia. Similar clinical and laboratory features were observed by
BIGLIERI et al. (1966b) in a patient with 17-hydroxylation deficiency
during treatment with low doses of dexamethasone. However,
in a second case of 17-hydroxylation deficiency, the decreased aldosterone secretion rate became normal when the excessive production of corticosterone and deoxycorticosterone was suppressed
by dexamethasone therapy (GOLDSMITH et al., 1967). According to BIGLIERI et al. (1966b), the persistently decreased aldosterone
secretion rate observed in their patient could have been due either
to chronic suppression by deoxycorticosterone or to a second bio-

---

4 The effect of serotonin and melatonin, two other constituents of
pineal gland extracts, on aldosterone biosynthesis is described in chapter
IV-7.

synthetic defect in the aldosterone pathway. Decreased aldosterone secretion was found in a patient with a deoxycorticosterone-producing, feminizing adrenocortical carcinoma, who was hypertensive but had normal serum electrolytes (SOLOMON et al., 1968). Aldosterone secretion rates or urinary excretion of aldosterone-18-glucuronide were found to be abnormally low in patients with the hypertensive form of congenital adrenal hyperplasia (11β-hydroxylation deficiency) with or without glucocorticosteroid therapy (NEW et al., 1966; KOWARSKI et al., 1968); this could have been due either to excessive deoxycorticosterone production or to deficiency of the 11β-hydroxylation step in the aldosterone biosynthetic pathway.

Administration of deoxycorticosterone acetate for several days or weeks to normal men (BIGLIERI et al., 1967) and rats (SINGER and STACK-DUNNE, 1955; EILERS and PETERSON, 1964) led to marked reductions in aldosterone secretion. Treatment of rats with 9α-fluorocortisol for two weeks resulted in a 70—90% decrease in aldosterone production *in vitro* by quartered adrenal glands incubated with or without aldosterone-stimulating substances but had no effect on corticosterone or deoxycorticosterone production (MÜLLER, 1970) (Fig. 18). Markedly reduced secretion rates of aldosterone were also observed in patients with excessive intake of licorice (JENNY et al., 1961; SALASSA et al., 1962; CONN et al., 1968). Licorice contains glycyrrhizic acid, which is not a steroid but a structurally related substance known to mimic the sodium-retaining and potassium-diuretic activity of aldosterone.

Aldosterone production by adrenal tissue of rats which were kept on a sodium-deficient diet during treatment with 9α-fluorocortisol was no different from aldosterone production by adrenals of untreated sodium-deficient rats under most conditions of incubation (MÜLLER, 1970) (Fig. 18). In sheep with a moderate sodium deficiency aldosterone secretion was not suppressed by adrenal arterial or systemic infusion of exogenous aldosterone (BLAIR-WEST et al., 1962, 1965b). These findings indicate that decreased aldosterone secretion is not due to a direct effect of plasma mineralocorticosteroids on the adrenal cortex or a receptor organ involved in negative feedback control, but is rather a consequence of a mineralocorticosteroid-induced alteration in sodium balance or sodium distribution. However, suppression of aldosterone secretion is not proportional to sodium retention. Thus, the same decreases in aldosterone excretion were induced by three days' treatment with deoxycorticosterone acetate in healthy men and in patients with essential hypertension (BIGLIERI et al., 1967). In the first group the cumulative sodium retention ranged between 250 and 750 mEq, weight gain between 2 and 4 kg; in the second group sodium retention was less than 150 mEq, weight gain less than 0.5 kg. No significant changes in serum electrolyte concentrations, blood pressure or hematocrit levels were observed in either group.

Decreases in aldosterone production due to exogenous or endogenous mineralocorticosteroids could be mediated by the renin-angiotensin system. Significant suppression of renin secretion was induced in dogs by treatment with aldosterone (GELHOUD and VANDER, 1967) and with deoxycorticosterone acetate (ROBB et al., 1969); this suppression was prevented by sodium deficiency. Decreased or low levels of plasma renin activity or plasma renin concentration are found in patients with primary aldosteronism (CONN et al., 1964; BROWN et al., 1968), 17-hydroxylation deficiency (BIGLIERI et al., 1966b; GOLDSMITH et al., 1967) or licorice-induced pseudoaldosteronism (CONN et al., 1968). However, following removal of an aldosterone-producing adenoma, aldosterone secretion may remain decreased at a time when plasma renin activity has returned to normal or is elevated (CONN et al., 1964). Prolonged administration of deoxycorticosterone acetate to rats resulted in marked narrowing and lipid depletion of the zona glomerulosa (DEANE et al., 1948). This effect was observed also in hypophysectomized rats (GREEP and DEANE, 1947) but not in animals kept on a sodium-deficient diet (DEANE et al., 1948). Thus, decreased aldosterone production may at least be partially due to atrophy of the zona glomerulosa. Little evidence is available as to which stages of aldosterone biosynthesis are particularly affected by mineralocorticosteroid excess. In adrenal regeneration hypertension, which is characterized by an increased production of deoxycorticosterone *in vitro* and *in vivo* (BROWNIE and SKELTON, 1965; RAPP, 1969b), VECSEI et al. (1966) found a decreased conversion to aldosterone of radioactively labelled progesterone and deoxycorticosterone added in trace amounts to quartered adrenals. However, specific activity was not measured. Moreover, alteration in aldosterone biosynthesis could have been due either to the adrenal lesion itself or to suppression by deoxycorticosterone.

6*

9. Glucocorticosteroids and ACTH

In short-term experiments, ACTH has been consistently found to stimulate aldosterone biosynthesis *in vitro* and *in vivo* (see chapter IV-4). However, its role in long-term regulation of aldosterone secretion has not yet been clearly established. Thus, it is not known at present whether increased plasma levels of ACTH substantially contribute to chronic increases in aldosterone secretion under certain physiological or pathological conditions and whether the functional capacity of the zona glomerulosa remains completely intact for prolonged periods of time in the absence of ACTH secretion.

Increased production of aldosterone and corticosterone in dogs with thoracic vena cava inferior constriction has been partially ascribed to increased ACTH secretion (DAVIS et al., 1960a,b; HOWARDS et al., 1968). In these animals, administration of high doses of cortisone as well as hypophysectomy resulted in marked decreases in aldosterone secretion. A few patients with hyperaldosteronism of unknown etiology have been observed in whom aldosterone secretion became normal under treatment with dexamethasone. Two of these cases (NEW and PETERSON, 1967; MIURA et al., 1968) had elevated plasma ACTH levels, whereas in the two cases (father and son) described by SUTHERLAND et al. (1966) plasma ACTH was normal. In all these cases, the increased secretion of aldosterone was possibly due to an abnormal response of the adrenal cortex to ACTH.

When presumably normal adrenal glands are chronically exposed to elevated levels of endogenous ACTH or to administration of exogenous ACTH, aldosterone secretion either remains normal or decreases. Thus, normal or decreased aldosterone secretion rates were observed in patients with Cushing's syndrome, due to increased ACTH secretion by the pituitary gland or by a non-endocrine tumor (BIGLIERI et al., 1968). Administration of high doses of ACTH to normal human subjects on a liberal or restricted sodium intake or to patients with essential hypertension or primary aldosteronism caused only transient increases in aldosterone excretion, which were followed by decreases below control values (NEWTON and LARAGH, 1968a; BIGLIERI et al., 1969). Administration of ACTH to rats for two weeks led to marked decreases of aldosterone production by adrenal tissue incubated with or without aldosterone-stimulating substances (MÜLLER, 1970) (Fig. 18). This suppression of aldosterone biosyn-

thesis was probably not due to an increased output of other mineralocorticosteroids, since ACTH administration to rats kept on a
sodium-deficient diet resulted also in a marked decrease in aldosterone production (see chapter V-8). On the other hand, the im-

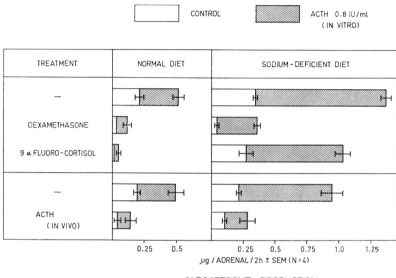

ALDOSTERONE PRODUCTION

Fig. 18. Aldosterone production *in vitro* by quartered adrenal glands of
rats receiving different diets and medications (dexamethasone and
9α-fluorocortisol added to the drinking fluid, 2 mg/l; ACTH-Zn injected
subcutaneously, 38 μg/d) for two weeks. White bars represent base-line
production, shaded bars increments due to ACTH added *in vitro*. Mean
values of two experiments (N = 4) ± standard error of the mean. (From
Müller, 1970)

pairment of aldosterone production could have been induced by the
augmented secretion of glucocorticosteroids. In sodium-deficient
men and rats, treatment with glucocorticosteroids resulted in marked
decreases in aldosterone secretion rate and in aldosterone production by adrenal tissue, respectively (Spark et al., 1968; Newton
and Laragh, 1968b; Müller, 1970; Fig. 18)[5]. Possibly, suppres-

---

5 On the other hand, Kolpakov et al. (1970) have found that administration of large doses of cortisol to dogs induced a significant increase
of aldosterone production *in vitro*.

sion of aldosterone production by glucocorticosteroids is independent of ACTH suppression. FEKETE and SZEBERENYI (1968) have shown that a number of glucocorticosteroids considerably differed in their aldosterone-suppressing activities when given to rats in doses which were approximately equivalent in suppressing corticosterone production. In human subjects cortisol suppressed aldosterone secretion more markedly than dexamethasone (NEWTON and LARAGH, 1968b).

The mechanism by which endogenous or exogenous glucocorticosteroids impair aldosterone biosynthesis is as yet unknown. Apparently, changes in sodium balance are not involved. Available evidence also seems to exclude mediation by the renin-angiotensin system. In man, suppression of aldosterone secretion by treatment with cortisol, dexamethasone or ACTH was not accompanied by consistent decreases in plasma renin activity (NEWTON and LARAGH, 1968a,b). Aldosterone secretion was suppressed in patients with primary aldosteronism in whom plasma renin activity was too low to be measured under control conditions. In the rat, short-term treatment by ACTH transiently stimulated and treatment with glucocorticosteroids inhibited renin release (HAUGER-KLEVENE et al., 1969). Angiotensin infusions did not restore aldosterone secretion rate to control levels in dexamethasone-treated men (SPARK et al., 1968). NEWTON and LARAGH (1968b) have suggested that inhibition of aldosterone biosynthesis could be due to a direct effect of glucocorticosteroids on the zona glomerulosa. As shown in chapter IV-10-a, glucocorticosteroids directly inhibit aldosterone biosynthesis *in vitro*, but only when they are added to the incubation medium in very high concentrations. Possibly, under long-term *in vivo* conditions lower concentrations of glucocorticosteroids could directly interfere with aldosterone biosynthesis. Inhibition could also be due to extra-adrenal metabolic effects of glucocorticosteroids. Glucocorticosteroids can elicit potassium losses by a mechanism which is independent of "aldosterone-like" activity (BARTTER and FOURMAN, 1957; LIDDLE, 1959). However, no significant changes in external potassium balance or serum potassium levels were observed by NEWTON and LARAGH (1968b).

There is very little evidence available to indicate at which site in the biosynthetic pathway aldosterone production is inhibited by glucocorticosteroid or ACTH administration. Aldosterone production of adrenal tissue of glucocorticosteroid-treated rats remained abnormally low when progesterone was added to the incubation medium in substrate amounts (FEKETE and SZEBERENYI, 1968). VECSEI et al. (1966) found a decreased incorporation of radioactive progesterone and deoxycorticosterone into aldosterone by adrenal tissue of ACTH-treated rats. The conversion of radioactive progesterone to aldosterone was impaired to a greater extent than the conversion of radioactive deoxycorticosterone to aldosterone.

10. Estrogens and Progestogens

a) Exogenous Estrogens

PETERSON et al. (1960) observed that treatment with ethinyl estradiol (0.5 mg/d) did not alter urinary aldosterone excretion in three normal subjects. According to LAIDLAW et al. (1962), long- or short-term treatment with diethylstilbestrol or ethinyl estradiol (various doses) led to moderate increases in aldosterone excretion; only in 1 out of 8 subjects was aldosterone excretion elevated to a pathological level. These investigators assumed that the increases in urinary aldosterone reflected rather altered aldosterone metabolism than increased secretion rates, since ethinyl estradiol (0.3 mg/d) also augmented aldosterone excretion in a patient with Addison's disease, who was treated with aldosterone infusions. However, LAYNE et al. (1962) found that treatment of 5 premenopausal women with ethinyl estradiol 3-methyl ether (0.3 mg/d) for 20 days resulted in a statistically significant increase of the mean aldosterone secretion rate from 76 to 151 µg/d, in a decrease of the fractional metabolism of tritiated aldosterone to glucuronidase-hydrolyzable urinary metabolites and in a markedly increased binding of aldosterone to plasma globulins from 16.4 to 40%. When the same estrogen was given at a lower dosage (0.1 mg/d), it had no effect on the secretion rate and plasma protein binding of aldosterone, but influenced aldosterone metabolism in a similar way as it did at the higher dosage. CRANE and HARRIS (1969a) treated 13 normal subjects with ethinyl estradiol (0.5 mg/d) for three weeks and observed a significant increase of the mean aldosterone excretion rate from 8.5 to 15 µg/d and from 29.3 to 61.2 µg/d on an unrestricted and on a low sodium intake, respectively. The increases in aldosterone excretion were accompanied by striking elevations in plasma renin activity as measured under three different combinations of sodium intake and body posture. The mechanism of these estrogen-induced increments in plasma renin activity and aldosterone secretion is unknown. Possibly they are secondary to increases in plasma renin substrate concentration. Treatment of rats with different estrogens and of two men with diethylstilbestrol (25 mg/d) resulted in marked increases in plasma renin substrate concentration (HELMER and GRIFFITH, 1952). Estrogens do not induce sodium diuresis as progesterone does but rather have a salt retaining activity. Thus, treatment with estradiol (10 mg/d) resulted in a decrease of NaCl and water excretion, which was small and transient in normal subjects but marked and sustained in patients with cirrhosis of the liver and ascites (PREEDY and AITKEN, 1956a,b).

b) Exogenous Progestogens

The natural hormone progesterone has a natriuretic activity, which is probably due to competitive inhibition of aldosterone action on the distal renal tubule. It is a potent inhibitor of aldosterone action on the toad bladder and displaces aldosterone from receptors in this tissue (for reviews see SHARP and LEAF, 1966). Treatment of subjects with normal adrenal function with progesterone (50—100 mg/d) resulted in marked but transient urinary sodium and chloride excretion. In mineralocorticoid-treated but not in untreated Addisonian patients, progesterone induced a pronounced sodium loss (LANDAU et al., 1955). In short-term experiments carried out in two patients with Addison's disease, 150 mg of progesterone completely blocked the sodium-retaining effect of 20 µg, but not of 40 or 50 µg of aldosterone (LANDAU and LUGIBIHL, 1958).

LAIDLAW et al. (1962) treated 6 subjects with progesterone (50 to 200 mg/d) and observed an increase in urinary aldosterone excretion in all instances. However, this increase was moderate in two elderly subjects (62 and 64 years), who also showed no reversal of progesterone-induced natriuresis. In 4 younger subjects, sodium diuresis was reversed within 3 days of progesterone administration, and urinary aldosterone excretion rose to values above the normal range. In two subjects aldosterone secretion rate was also determined; it rose in proportion to rises in aldosterone excretion. On the other hand, increases in aldosterone excretion induced by progesterone (100—200 mg/d) treatment in 9 women were very small, and abnormally high values were observed only in three instances (STARK and KOSSMANN 1963). In these patients, sodium diuresis was either absent or moderate. LAYNE et al. (1962) induced significant increases in aldosterone secretion rate in 5 women by oral treatment with progesterone (300 mg/d); this medication had no effect on aldosterone metabolism. Administration of progesterone (200 mg/d) significantly enhanced aldosterone secretion rate in women with normal pregnancy as well as in patients with a pregnancy terminated by intrauterine fetal death; in several instances aldosterone secretion did not return to control values but remained elevated or increased even further upon cessation of progesterone treatment (WATANABE et al., 1965). Treatment of rats with progesterone (10 mg/d) for 4—6 days resulted in a 100% increase of aldosterone secretion rate and in a smaller but significant decrease in corticosterone production (SINGER et al., 1963b; TELEGDY and LISSAC, 1965). In hypophysectomized rats, exogenous progesterone stimulated aldosterone secretion to the same extent it did in intact animals, but had no effect on corticosterone output (SINGER et al., 1963b).

The natriuretic activity of progesterone appears to be independent of its progestational activity, since synthetic steroids with potent progestational activity were found either to have no effect on sodium excretion or to lead to sodium retention (LANDAU et al., 1958; CHAGOYA et al., 1961). Treatment with norethinodrel (17-ethinyl-17β-hydroxy-5(10)-estrene-3-one, 10 mg/d) induced a small but statistically significant elevation of aldosterone secretion rate in premenopausal women (LAYNE et al., 1962). On the other hand, administration of medroxyprogesterone acetate (6α-methyl-17α-hydroxy-4-pregnene-3, 20-dione acetate; 10 or 20 mg/d) for three weeks led to a marked reduction of the aldosterone excretion rate in 5 normal subjects on an unrestricted sodium intake, but did not alter the response in aldosterone excretion to dietary sodium restriction (CRANE and HARRIS, 1969a). This indicated that the suppression of aldosterone secretion was due to sodium-retaining activity of the drug. However, suppression of aldosterone secretion was not accompanied by a decrease in plasma renin activity.

c) Oral Contraceptive Medication

Since oral contraceptives generally contain an estrogenic as well as a progestational component, they could theoretically exert a combination of estrogen-like and progesterone-like effects on aldosterone secretion and metabolism. However, the dose of estrogen in most of these preparations is lower than the minimum dose needed to induce a predictable effect on aldosterone secretion and plasma renin activity in normal human subjects. The progestational component of oral contraceptives is mainly of the type that either does not influence sodium excretion or has a transient sodium-retaining activity.

The incidence of increased aldosterone secretion rate among five women taking Enovid[6] (10 mg/d) was 100% according to LAYNE et al. (1962). In these women aldosterone metabolism was altered in the same manner as it was in women taking ethinyl estradiol 3-methyl ether (0.3 mg/d), and aldosterone binding to plasma globulins was markedly increased (MEYER et al., 1961). A much lower incidence of increased aldosterone secretion or excretion rate among women on oral contraceptive medication has been reported by LARAGH et al. (1967) and CRANE and HARRIS (1969b). Three out of 5 women who were treated with Enovid[6] (5 mg/d) during one cycle showed a marked increase of the aldosterone secretion rate on the 7th day of the Enovid cycle compared to their control cycle (GRAY et al., 1968). On the 19th and 25th day of the Enovid cycle aldosterone secretion rate was in most instances lower than it was on the respective days of the control cycle.

A marked elevation in the plasma concentration of renin substrate to a level characteristic of normal pregnancy has been found in almost all women on oral contraceptive medication (HELMER and JUDSON, 1967; LARAGH et al., 1967). Plasma renin activity was less frequently and less markedly elevated. Plasma renin concentration was normal in 6 women taking oral contraceptives (HELMER and JUDSON, 1967).

d) Menstrual Cycle

The mean aldosterone secretion rate of 13 normal women was 139 μg/d during the follicular phase of the menstrual cycle (GRAY

---

6 98.5% 17-ethinyl-17β-hydroxy-5(10)-estrene-3-one and 1.5% ethinylestradiol 3-methyl ether.

et al., 1968). This value corresponded well to the mean aldosterone secretion rate of normal men (135 µg/d, WATANABE et al., 1963). During the luteal phase, there was a significant rise in aldosterone secretion to a mean value of 235 µg/d. This increase was also observed in women on a high sodium intake. The rise in aldosterone secretion was associated with an increase in urinary pregnanediol excretion, but was not accompanied by a consistent alteration in sodium balance. Fluctuations of aldosterone output during the menstrual cycle could be mediated by the renin-angiotensin system, since a small but significant rise in plasma renin concentration was observed in normal women during the 3rd or 4th week, i.e. the luteal phase, of the menstrual cycle (BROWN et al., 1964b).

e) Pregnancy

An increased secretion rate of aldosterone is commonly found in the third trimester of normal pregnancy. A mean aldosterone secretion of 582 µg/d was found in 6 women in the 32nd to 38th weeks of pregnancy (JONES et al., 1959). Similar values have been observed by STARK (1966). More pronounced increases in aldosterone secretion have been reported by WATANABE et al. (1963). These investigators found aldosterone secretion to be elevated in women who were in the 15th and 20th weeks of pregnancy. Between the 28th and 35th weeks of pregnancy aldosterone secretion rate was relatively constant in individual subjects; the mean levels were around 1000 µg/d. Between the 36th week and delivery there was a further increase to a mean value of 1600 µg/d. Aldosterone secretion rate was inversely related to sodium intake. On a sodium intake of less than 10 mEq/d aldosterone secretion rose to values as high as 7800 µg/d.

The physiological importance of the enhancement of aldosterone secretion during late pregnancy is unknown. The aldosterone production has its origin in the maternal adrenals. Aldosterone was not detected in the urine of a pregnant woman with Addison's disease (BAULIEU et al., 1957). Extremely low values of urinary aldosterone excretion were observed in two bilaterally adrenalectomized women during pregnancy (LAIDLAW et al., 1958). Increased aldosterone secretion depends on a living fetus; after intrauterine fetal death aldosterone secretion rate fell to levels found in non-pregnant women (WATANABE et al., 1965). Increased aldosterone production does not reflect increased aldosterone catabolism; the metabolic clearance rate of aldosterone was found to be the same in non-pregnant and pregnant women (TAIT et al., 1962). LAIDLAW et al. (1958) treated two adrenalectomized women with a constant maintenance dose of cortisol and 9α-fluorocortisol throughout pregnancy and concluded that "the high output of aldosterone during pregnancy was not a requisite for healthy gestation and served no apparent purpose". However, according to VENNING et al. (1959) increased aldosterone secretion may contribute to the substantial sodium retention necessary for uterine and fetal development and for the expansion of maternal blood volume during normal pregnancy. Enhanced aldosterone production may also be necessary to overcome the aldosterone-antagonistic activity of the high levels of circulating progesterone in pregnant women. Hypokalemia, metabolic alkalosis and hypertension markedly improved during pregnancy in a woman with primary aldosteronism, despite a large increase in aldosterone secretion (BIGLIERI and SLATON, 1967). A direct correlation between aldosterone secretion rate and urinary pregnanediol excretion has been observed by JONES et al. (1959). EHRLICH et al. (1962) observed that in three pregnant women an increased sodium intake resulted not only in suppression of aldosterone secretion but also in a concomitant decrease in pregnanediol excretion.

The role of the renin-angiotensin system in mediating a high aldosterone secretion rate in pregnancy is uncertain. Elevation of plasma renin is found in approximately half of all pregnant women examined, whereas the other half have renin concentrations in the upper range of normal non-pregnant subjects (BROWN et al., 1963; GOULD et al., 1966; HELMER and JUDSON, 1967). There is no apparent correlation between plasma renin and duration of pregnancy. Some of the highest values have been observed in the first trimester. The origin of the renin found in the peripheral plasma of pregnant women is unknown. High concentrations of the enzyme have been found in human amniotic fluid (BROWN et al., 1964c) and in the uterus and placenta of rabbits (GROSS et al., 1964). Abnormally low values of plasma renin activity were observed in a pregnant woman with primary aldosteronism (GORDON et al., 1967). A marked elevation of plasma renin substrate concentration is a consistent finding in human pregnancy (GOULD et al., 1966; HELMER and JUDSON, 1967). In contrast, plasma renin substrate is significantly decreased during pregnancy in the rat (CARRETERO and GROSS, 1967).

Decreased aldosterone secretion rates have been observed in women with severe pre-eclampsia (WATANABE et al., 1965). This corresponds to the decreased plasma renin levels found in pregnant women with hypertension and proteinuria (BROWN et al., 1965b, 1966b). These findings may indicate pathological sodium retention in this disease. KUMAR et al. (1959) found an increased intracellular sodium concentration in pre-eclamptic patients, whereas intracellular sodium was decreased in normal late pregnancy.

11. Heparin and Heparinoids

The natriuretic effect of heparin and of heparinoids[7] has been found to be at least mainly due to suppression of aldosterone secretion. These drugs markedly lower the urinary excretion and the secretion rate of aldosterone in normal men on a liberal or restricted sodium intake and in patients with essential hypertension or edema (CEJKA et al., 1960; VEYRAT et al., 1963; SCHLATMANN et al., 1964; ABBOTT et al., 1966b). Heparin and heparinoids also decrease aldosterone secretion in patients with primary aldosteronism (FORD and BAILEY, 1966; CONN et al., 1966; ABBOTT et al., 1966b), i.e. in patients with a very low plasma renin activity; they do not affect 17-hydroxycorticosteroid or 17-ketogenic steroid excretion, and they lead to increases in the plasma potassium concentration. These findings indicate that inhibition of aldosterone secretion by these drugs is not due to inactivation of the renin-angiotensin system, suppression of ACTH secretion or hypokalemia. When added to the medium in which rat adrenal tissue was incubated, heparin or heparinoids did not suppress *in vitro* production of aldosterone (GLAZ and SUGAR, 1964). This has been confirmed by studies in our laboratory; at a concentration of 0.5 mg/ml, Ro-1-8307 affected neither

7 Although a number of different heparin derivatives exert a similar activity, the heparinoid most frequently used in these investigations has been Ro 1-8307 (N-formyl chitosan polysulphuric acid), which has a very low anticoagulant activity but is almost as active as heparin in promoting natriuresis.

aldosterone production by rat adrenal incubates nor its stimulation by a high potassium concentration (MÜLLER, unpubl.). Significantly decreased amounts of aldosterone were produced *in vitro* by adrenals taken from rats 6 hours after injection of a single dose of heparin or heparinoids *in vivo* (GLAZ and SUGAR, 1964). Adrenals of rats treated with Ro 1-8307 (20 mg/d) for three days produced normal amounts of aldosterone under basal conditions of incubation; when serotonin or KCl was added to the incubation medium, adrenals of heparinoid-treated rats produced approximately 30% less aldosterone but 30% more corticosterone and deoxycorticosterone than adrenals of control animals (MÜLLER, unpubl.). Heparinoid treatment decreased the secretion rate of 18-hydroxycorticosterone (ABBOTT et al., 1966b) and the urinary excretion of the major metabolite of 18-hydroxycorticosterone (CONN et al., 1966), but did not alter the excretion of tetrahydrocorticosterone. These data indicate that the inhibition of aldosterone biosynthesis occurs at the 18-hydroxylation step. Prolonged treatment with heparin of a patient with idiopathic hypercholesterolemia resulted in marked atrophy of the zona glomerulosa of the adrenal cortex and an increased juxtaglomerular cell index in the kidney (WILSON and GOETZ, 1964). Long-term administration of Ro 1-8307 to rats led to atrophy of the zona glomerulosa (ABBOTT et al., 1966a).

VI. Conclusions

1. Multiplicity of Aldosterone-Stimulating Substances

Several substances directly stimulate aldosterone production by isolated rat adrenal tissue during two hours of incubation in a medium containing only normal plasma electrolytes and glucose: these are angiotensin II, potassium ions, ammonium ions, rubidium ions, caesium ions, ACTH, cyclic AMP, serotonin and some related compounds[1]. With the exception of angiotensin II, which even in very high doses stimulated aldosterone production to a relatively small extent, these substances reproducibly stimulated aldosterone production by adrenal tissue of sodium-deficient rats by 100 to 400% above mean control values. Only potassium ions stimulated aldosterone production when added to the incubation medium in a concentration which corresponded to a physiological plasma level. Serotonin significantly augmented aldosterone biosynthesis at a very low dose of 10^{-8} moles/l; however, it is uncertain whether this dose is a physiological one, since most if not all circulating serotonin is stored by the blood platelets under physiological conditions (RAND and REID, 1951; HUMPHREY and TOH, 1954). The other substances stimulated aldosterone production at rather high concentrations. This does not imply that their aldosterone-stimulating activity is not of potential physiological importance. ACTH, which is the major biological regulator of glucocorticosteroid production, also stimulated the *in vitro* production of corticosterone by rat adrenal tissue only when added to the incubation medium in a dose far above the physiological range.

Each of these aldosterone-stimulating substances acted independently of other stimulators, but all of them presumably acted on the same cells. Potassium ions stimulated aldosterone production maxi-

---

1 The aldosterone-stimulating effect of added NADP and glucose-6-phosphate will not be mentioned in this context, since it may depend on damaged cells and thereby represent an artefact of the *in vitro* system (HALKERSTON et al. 1968).

mally in a serotonin- and ACTH-free medium. ACTH and serotonin, respectively, induced a maximal increase in aldosterone output when added to an incubation medium with a potassium concentration of 3.6 mEq/l, which is at the lower limit of the normal range
of plasma potassium levels. Serotonin stimulated aldosterone production in a potassium-free medium. When two aldosterone-stimulating agents were simultaneously added to the incubation medium
— each at a dose which by itself maximally stimulated aldosterone
production — no cumulation of the steroidogenic effects was observed. The principle of multiple and independent aldosterone-stimulating agents that has been observed *in vitro*, appears to pertain to
short-term stimulation of aldosterone secretion *in vivo*, too. Thus,
angiotensin stimulated aldosterone production in hypophysectomized dogs (DAVIS, 1962), and potassium is a potent aldosterone
stimulator in nephrectomized and hypophysectomized animals
(DAVIS et al., 1963). Also, in certain cases of renovascular hypertension a high output of aldosterone is maintained presumably by
the renin-angiotensin system despite marked hypokalemia (see chapter V-4-a). This apparent equality of aldosterone-stimulating
substances sharply contrasts with the exclusive role of ACTH in
the regulation of glucocorticosteroid secretion. In a hypophysectomized animal no known humoral agent, with the possible exception of cyclic AMP, can fully replace the cortisol- and corticosterone-
stimulating effect of ACTH.

If we accept the principle of multiple and equal aldosterone-stimulating substances, we must also accept the possibility that some potent aldosterone-stimulating agents are still undiscovered. Such
factors could be of clinical importance. Several investigators have
suggested that "non-tumorous primary aldosteronism", i.e. hypersecretion of aldosterone, decreased plasma renin activity, hypertension, hypokalemia and metabolic alkalosis without an adrenal
tumour but with bilateral nodular adrenocortical hyperplasia, could
be due to a yet unknown aldosterone-stimulating principle (ROSS,
1965; DAVIS et al., 1967; KATZ, 1967; SALTI et al., 1969).

2. "Adrenoglomerulotropins"

The term "adrenoglomerulotropin" was introduced by FARRELL
(1960), who used it to designate a hypothetical hormone of dience-

phalic or pineal origin which controlled specifically the secretion of aldosterone by the zona glomerulosa of the adrenal cortex. Later, evidence was presented by FARRELL and McISAAC (1961) which indicated that "adrenoglomerulotropin" was identical with 6-methoxytetrahydroharman, i.e. a carboline derivative. This substance did not affect aldosterone production by sheep or dog adrenals *in vivo* or by rat adrenals *in vitro*. However, the term "adrenoglomerulotropins" would very appropriately describe the biological activity of a group of other substances which do stimulate aldosterone production. This group of "adrenoglomerulotropins" includes potassium ions, related monovalent cations, angiotensin II, serotonin and related indole derivatives. "Adrenoglomerulotropins" stimulate steroidogenesis exclusively or preferentially in the zona glomerulosa, but they do not stimulate aldosterone production under all experimental conditions and they also stimulate the biosynthesis of other steroids produced by the zona glomerulosa.

In adrenals of sodium-deficient rats "adrenoglomerulotropins" stimulated mainly aldosterone production and did not have a significant effect on corticosterone production, in contrast to ACTH and cyclic AMP which stimulated aldosterone and corticosterone production to a similar extent. Tracer studies indicated that during short-term incubations "adrenoglomerulotropins" stimulated aldosterone biosynthesis by enhancing the conversion of cholesterol to pregnenolone. Since these reactions are common to the synthesis of all steroid hormones and are also the main site of action of ACTH, I have suggested (MÜLLER, 1966) that the preferential stimulation of aldosterone biosynthesis by certain compounds could be due to a selective action on cells that produce mainly aldosterone, i.e. zona glomerulosa cells. Recent incubation studies with separate zones of rat adrenals have confirmed this hypothesis (MÜLLER, 1971). Serotonin, potassium ions and angiotensin stimulated steroidogenesis only in the capsular portion (zona glomerulosa) of rat adrenals but had no effect on the steroid production of the decapsulated portion (zona fasciculata and reticularis). ACTH and cyclic AMP stimulated steroidogenesis to a similar extent in both adrenal capsules and decapsulated adrenals. In adrenals of potassium-deficient rats, potassium ions and serotonin did not stimulate aldosterone production, but significantly augmented the production of deoxycorticosterone. In these adrenals, also, the response in deoxycorticosterone output to stimulation by serotonin and potassium ions is completely due to production by the capsular portion. Thus, at least in short-term incubation experiments with rat adrenals, serotonin, potassium ions and angiotensin are true "adrenoglomerulotropins" stimulating steroidogenesis exclusively in the zona glomerulosa even in adrenals in which they do not stimulate aldosterone production.

In other species such as sheep, dogs and cattle, angiotensin appears to have at least a preferential action on the zona glomerulosa. In hypophysectomized dogs high doses of angiotensin stimulate cortisol secretion. However, in such animals sodium deficiency also leads to increased cortisol secretion, which may indicate that some cortisol is produced by the zona glomerulosa of the dog adrenal. On the other hand, there is no convincing evidence that potassium ions can directly stimulate cortisol biosynthesis *in vivo* or *in vitro* in any animals species, with the possible exception of the dog (BURWELL et al., 1969.)

3. Uniform Mode of Action of Aldosterone-Stimulating Compounds

As shown in the preceding chapter, certain aldosterone-stimulating substances, such as serotonin, monovalent cations and angiotensin, act only on zona glomerulosa cells whereas ACTH also acts on zona fasciculata-reticularis cells. According to available evidence it is possible that "adrenoglomerulotropins" and ACTH enhance steroidogenesis in their respective target cells by a similar biochemical mechanism. After a decade of intensive research by a large number of very competent investigators, the exact sequence of complex events by which ACTH stimulates steroidogenesis in the adrenocortical cells is still unknown and controversial (HILF, 1965; HALKERSTON, 1968; KORITZ, 1968; McKERNS, 1968). I shall therefore not presume to propose a complete theory on the mechanism of action of aldosterone-stimulating substances on the zona glomerulosa cell on the basis of just a few observations.

a) Site of Action in the Biosynthetic Pathway

As previously stated, the assumption that aldosterone-stimulating substances act on the conversion of cholesterol to pregnenolone and do not influence later biosynthetic steps during short-term incubation experiments is primarily based on the observation that potassium ions, ammonium ions, serotonin and angiotensin did not affect the incorporation of added tritiated pregnenolone, progesterone, deoxycorticosterone or corticosterone into aldosterone, but stimulated the conversion of tritiated cholesterol to aldosterone. Supporting evidence for a site of action early in the biosynthetic pathway can be taken from the following observations:

1. Direct stimulation of aldosterone production by angiotensin, potassium ions or serotonin *in vitro* or *in vivo* was never accompanied by a concomitant decrease in the output of aldosterone-precursor steroids such as corticosterone or deoxycorticosterone. The production of these steroids was either unaffected or stimulated by "adreno glomerulotropins".

2. When the activity of the enzymes involved in the final stages of aldosterone biosynthesis was decreased by potassium deficiency or was blocked by Su 8000, potassium ions or serotonin induced an increment in deoxycorticosterone production which was similar to the increment in aldosterone production induced by these stimulators under other experimental conditions.

3. Serotonin did not stimulate aldosterone production in adrenal tissue maximally stimulated by ACTH. The main site of action of ACTH is also the conversion of cholesterol to pregnenolone in the corticosteroid biosynthetic pathway during short-term stimulation (STONE and HECHTER, 1954; KARABOYAS and KORITZ, 1965).

4. Angiotensin did not stimulate aldosterone production by beef adrenal slices, when substrate amounts of progesterone or corticosterone were added to the incubation medium (KAPLAN and BARTTER, 1962).

5. Angiotensin did not increase the conversion of tritiated corticosterone to aldosterone by perfused sheep adrenals (BLAIR-WEST et al., 1967).

b) Action on the Cell Membrane

There is some indirect evidence indicating that aldosterone-stimulating substances act primarily on the cell membrane:

1. Angiotensin did not stimulate aldosterone production by beef adrenal homogenates (KAHNT and NEHER, 1965).

2. The aldosterone-stimulating activity of potassium ions and of angiotensin was dependent on the presence of calcium ions in the incubation medium.

3. Ouabain, an agent which blocks the sodium-potassium pump of the cell membrane (SCHATZMANN, 1953), completely or partially inhibited the aldosterone-stimulating effect of potassium ions, serotonin and ACTH.

4. Cyclic AMP stimulated aldosterone production by rat or beef adrenal tissue. On incubates of capsular rat adrenal glands it induced a steroidogenic response similar to the one induced by serotonin, potassium ions or ACTH. Thus it is possible that aldosterone-stimulating agents—in analogy to the mode of action of other hormones (SUTHERLAND et al., 1965)—could activate adenyl cyclase, which is located in the cell membrane, and that cyclic AMP could act as a second messenger mediating steroidogenesis within the zona glomerulosa cell.

5. The onset of the steroidogenic response of the adrenal cortex to stimulation by angiotensin II and potassium ions, respectively, is very rapid. In perfused sheep adrenals a maximum in aldosterone output was elicited by these agents within 10—15 minutes (BLAIR-WEST et al., 1970a).

c) Dependence on Protein Synthesis

Cycloheximide completely inhibited the stimulation of aldosterone production by serotonin or potassium ions. This indicates that the response of the zona glomerulosa cells to these agents depends on the integrity of protein synthesis in a similar manner as the steroidogenic response of the zona fasciculata cells to ACTH does (HILF, 1965; FERGUSON, 1968; FARESE, 1968). However, as demonstrated by BURROW et al. (1966), inhibition of protein synthesis may also interfere with aldosterone production at late steps of the biosynthetic pathway, which are not directly affected by "adrenoglomerulotropins".

4. Regulation of the Final Steps of Aldosterone Biosynthesis

Experimental evidence indicates that alterations in sodium balance lead to considerable changes in the activity of one or both of the enzymes involved in the conversion of corticosterone to aldosterone, i.e. 18-hydroxylase and18- hydroxydehydrogenase. In addition, the activity of the final steps of aldosterone biosynthesis was found to be diminished by potassium deficiency and to be enhanced by uremia following bilateral nephrectomy. According to indirect evidence obtained in our laboratory, alterations in the activity of the conversion of corticosterone to aldosterone were accompanied by parallel alterations in the activity of the 11β-hydroxylation step in the aldosterone biosynthetic pathway. Alterations in the activity of 11β-hydroxylation and of 18-hydroxylation due to changes in sodium and potassium balance appear to be restricted to the zona glomerulosa of the adrenal cortex.

Increased conversion of endogenous or added corticosterone to aldosterone due to sodium deficiency has been observed in sheep adrenals perfused *in vivo* as well as in dog and rat adrenal tissue incubated *in vitro*. Moreover, it was demonstrable in a mitochondrial preparation of rat adrenal tissue incubated with various amounts of substrate, in four different buffers and in the presence of presumably optimal amounts of NADPH (MARUSIC and MULROW, 1967b). These latter experiments indicated that the increased conversion of corticosterone to aldosterone reflected an increase of enzyme activity, which was independent of the actual intracellular electrolyte composition and of the presence of aldosterone-stimulating substances and which was probably not due to increased availability of coenzyme.

Increases in the activity of the final steps of aldosterone biosynthesis due to sodium deficiency or uremia persisted during *in vitro* incubation, but they have as yet only been induced *in vivo*[2]. Preincubation of rat adrenal tissue for two hours in a medium containing serotonin, a high potassium concentration, or serum of sodium-deficient rats, did not result in an increased activity of the final steps of aldosterone biosynthesis (MÜLLER, unpubl.). Thus, available evidence does not allow any firm conclusions as to the mechanisms by which the activity of the final steps of aldosterone biosynthesis is altered in response to changes in sodium and potassium balance.

---

2 Decreases in the activity of the final steps of aldosterone biosynthesis can be directly induced *in vitro* by pharmacological agents such as Su 8000 or puromycin.

Perhaps, these alterations in enzyme activity are due to a long-term effect of blood electrolytes on the zona glomerulosa. However, they could also be mediated by a unknown extra-adrenal humoral stimulator or inhibitor.

Some evidence indicates that these alterations in the activity of the final steps of aldosterone biosynthesis occur rather slowly over periods of days, whereas activation of early steps by known aldo-sterone-stimulating substances occurs within minutes. Significant activation of the conversion of corticosterone to aldosterone has been observed after one day, maximum activation after two days of sodium deprivation in rats. Similary we have observed that addition of potassium chloride to the drinking fluid of rats for two days completely corrected the inactivation of the final steps of aldo-sterone biosynthesis induced by two weeks of dietary potassium deficiency (MÜLLER, unpubl.). However, in the sheep, increased conversion of tritiated corticosterone to aldosterone was observed only during the early stages of sodium deficiency.

Activation of the final steps of aldosterone biosynthesis can occur in the complete absence of the renin-angiotensin system. This has been demonstrated in rats with uremia due to bilateral nephrectomy. Experimental renal hypertension of rats induced by unilateral renal artery constriction resulted in marked increases in aldosterone pro-duction—presumably mediated by the renin-angiotensin system—which appeared to be mainly due to activation of early biosynthetic steps. These findings do not indicate that the marked activation of the final steps of aldosterone biosynthesis observed in sodium-deficient rats is mediated by the renin-angiotensin system.

Since all known aldosterone-stimulating substances have been shown to act on early steps of aldosterone biosynthesis, the theo-retical importance of a control system regulating the final steps is obvious. As has been shown by in vitro studies with adrenal tissue of potassium-deficient rats, inactivation of the final steps of aldosterone biosynthesis can completely block the "aldosterone-stimulating" (but not the steroidogenic) effect of "adrenoglomerulotropins" and of ACTH. On the other hand, activation of the final steps greatly increases the response of the adrenal cortex in aldosterone output to stimulation by these agents, as has been demonstrated in adrenal tissue of uremic and sodium-deficient rats.

5. Zona Glomerulosa Width

Elevated secretion of aldosterone often is associated with a wide zona glomerulosa, whereas a narrow zona glomerulosa often is found in animals with a diminished secretion of aldosterone. However, the morphological parameter of zona glomerulosa thickness is not a reliable index of the adrenal's capacity to produce aldosterone *in vivo* or *in vitro*. Thus, marked increases in the zona glomerulosa width were induced by sodium deficiency in hypophysectomized rats, in which aldosterone secretion was not stimulated (PALMORE and MULROW, 1967). Increased aldosterone production by rat adrenal tissue due to uremia following bilateral nephrectomy was not accompanied by a significant increase in zona glomerulosa thickness (MÜLLER and HUBER, 1969).

Zona glomerulosa thickness does not reliably reflect the adrenal's capacity to convert corticosterone to aldosterone. Addition of sodium chloride to the drinking fluid of rats for two weeks resulted in a marked decrease of the conversion of corticosterone to aldosterone by incubated adrenal tissue (MÜLLER, 1968), but in only negligible morphological alteration of the zona glomerulosa (HUBER, unpublished observation).

Zona glomerulosa width does not parallel the total increase in steroid production of the adrenal cortex in response to stimulation by "adrenoglomerulotropins". The response to serotonin in production of corticosterone and deoxycorticosterone of adrenal tissue taken from potassium-deficient rats was greater than the response in corticosterone, deoxycorticosterone and aldosterone production of adrenal tissue of sodium-deficient rats. The mean zona glomerulosa width was 24 μ in potassium-deficient and 59 μ in sodium-deficient rats (MÜLLER and HUBER, 1969).

Available evidence is at least compatible with the assumption that, in the rat, an alteration in the activity of the renin-angiotensin system generally leads to a corresponding alteration in zona glomerulosa width.

6. Species Differences

The relative or absolute lack of responsiveness of the rat adrenal cortex to stimulation by exogenous angiotensin or renin has been thought by some investigators to be indicative of a fundamental difference between the rat and other animal species in the physiological control of aldosterone secretion. However, indirect evidence indicates that in the rat as well as in other animal species there is a close functional relation between the renin-angiotensin system and the zona glomerulosa of the adrenal cortex (GROSS et al., 1965; GROSS, 1967). Also, there are considerable qualitative and quantitative species differences in almost any aspect of humoral control of aldosterone secretion in man, the dog, the sheep and the rat. The human and sheep adrenals also differ from the dog adrenal in their response to stimulation by angiotensin, since their aldosterone production is stimulated only by pressor doses of the peptide. Moreover, in the sheep, aldosterone secretion was not sustained at a high rate for more than a few hours by continous infusion of angiotensin as it was in the dog and in man. Angiotensin and sodium deficiency elicited increases in cortisol secretion in hypophysectomized dogs but not in sheep and man. Sodium deficiency increases the sensitivity of the adrenal cortex, to the aldosterone-stimulating effect of angiotensin in dogs, decreases it in man and practically blocks it in the sheep.

Marked differences between animals species exist also in the role of the kidneys and the pituitary in the control of aldosterone secretion. Nephrectomy immediately decreases the elevated secretion rate of aldosterone in sodium-deficient dogs but not in sodium-deficient sheep and rats. Hypophysectomy in sheep and dogs did not prevent a marked increase in aldosterone secretion in response to sodium deficiency as it did in the rat. The rat adrenal also appears to be more sensitive to the aldosterone-suppressing effects of long-term glucocorticosteroid administration than the dog or the human adrenal.

7. Predominance of the Renin-Angiotensin System in the Physiological Control of Aldosterone Secretion?

The possibility that the renin-angiotensin system plays an imimportant part in the regulation of aldosterone secretion was first suggested by GROSS (1958), who had observed an inverse relationship between sodium balance on one side and aldosterone secretion and renin content of the kidneys on the other side. In a later review, GROSS (1967) has warned that the term "aldosterone-stimulating hormone" in defining the physiological role of angiotensin II was "an oversimplification, as it might create the incorrect impression that angiotensin has an exclusive action on and elicits an exclusive response from the aldosterone-producing cells of the adrenal cortex". Although it has been acknowledged that plasma sodium and potassium concentrations as well as ACTH can also directly influence aldosterone secretion, the renin-angiotensin system is generally considered to be the most important factor in the overall physiological control of aldosterone output, which is reflected by the frequent use of the term "renin-angiotensin-aldosterone system". Often, alterations in aldosterone secretion which are not associated with changes in the plasma electrolyte levels or which are not accompanied by a significant alteration in glucocorticosteroid secretion are ascribed to mediation by the renin-angiotensin system.

To my knowledge, the only physiological situation in which a direct causal relation between the renin-angiotensin system and aldosterone secretion has been convincingly established is the upright posture of man. Sitting or standing has been found to induce parallel increases in urinary excretion (MULLER, 1963; WOLFE et al., 1966), plasma concentration and blood production rate of aldosterone (BALIKIAN et al., 1968) and plasma renin activity (GORDON et al., 1966) or concentration (BROWN et al., 1966c). Moreover, a decreased response in aldosterone production to postural changes was observed in patients with primary aldosteronism, and no response occurred in an anephric patient (BALIKIAN et al., 1968).

In a second physiological situation, i.e. in normal human pregnancy, increased secretion of aldosterone is associated with increased plasma renin and renin-substrate concentrations. However, aldosterone secretion rate was found to be markedly increased during the last trimester of pregnancy in all subjects studied and to reach its highest value in the last weeks before delivery. On the other hand, plasma renin concentrations were within the normal range in approximately half of normal women throughout pregnancy, and there was no correlation between the duration of pregnancy and plasma levels of renin.

Increased activity of the renin-angiotensin system considerably contributes to hypersecretion of aldosterone secondary to pathological conditions such as hemorrhage, constriction of the thoracic inferior vena cava and certain types of renovascular hypertension.

This has been demonstrated by simultaneous increases in aldo-sterone secretion and plasma renin and by the marked decreases in aldosterone secretion following nephrectomy. In addition, exoge-nous renin and angiotensin can raise the aldosterone secretion rate in normal animals to levels observed in animals with these various types of experimental secondary hyperaldosteronism. We can also assume that—at least in dogs—maintenance of a high aldosterone secretion rate in severe sodium depletion depends to a large extent on elevated plasma renin levels.

Thus, according to all the experimental evidence cited in this review, the renin-angiotensin system is a potent stimulator of aldo-sterone secretion in a number of pathological situations as well as in orthostasis of man. Many investigators conclude by extrapolation of this evidence that a normal aldosterone secretion rate is depen-dent on a normal plasma renin level and that decreased secretion of aldosterone is generally secondary to a decreased plasma renin concentration. The latter assumption is supported by the fact that sodium loading or administration of mineralocorticosteroids lead to parallel decreases in aldosterone secretion and plasma renin. However, there is no convincing evidence demonstrating that sub-normal levels of renin *per se* cause subnormal secretion of aldo-sterone. On the contrary, there is at least some evidence indicating that aldosterone secretion can be maintained at a normal basal rate despite very low plasma renin levels:

1. A normal aldosterone production rate was found in an anephric man (BALIKIAN et al., 1968).

2. Among patients with essential hypertension there is a high percentage of subjects with abnormally low plasma renin values not due to primary aldosteronism. A normal aldosterone secretion rate was found in most of these subjects when they were on a normal sodium intake (FISHMAN et al., 1968; HELMER and JUDSON, 1968; CHANNICK et al., 1969; JOSE and KAPLAN, 1969).

3. Bilateral nephrectomy did not result in decreases in aldo-sterone secretion in sodium-replete sheep (BLAIR-WEST et al., 1968b) and only in transient decreases in aldosterone secretion in intact dogs (GANN and TRAVIS, 1964).

4. Passive transfer of renin antibodies did not lower aldosterone secretion rate in sodium-replete dogs (LEE et al., 1965).

5. Administration of potassium to sodium- and potassium-depleted men resulted in an increased aldosterone secretion and in decreased plasma renin activity (VEYRAT et al., 1967).

In addition, increases in plasma renin or angiotensin II are not invariably followed by increases in aldosterone secretion:

1. According to WEIDMANN and SIEGENTHALER (1967), aldosterone secretion rate was abnormally high only in 2 out of 8 patients with renovascular hypertension and markedly elevated peripheral plasma renin activity. High plasma renin activity and normal aldosterone secretion rate was also found in patients with renal artery compression by tumors (WEIDMANN et al., 1969).

2. In some women with hypertension and increased plasma renin activity induced by oral contraceptives, aldosterone excretion rate was normal (LARAGH et al., 1967).

3. In two patients with diabetes insipidus, a very high level of plasma renin activity was accompanied by a normal and a pathologically decreased aldosterone excretion rate, respectively (WERNING et al., 1969).

4. In hypophysectomized dogs, a high sodium intake prevented a sustained increase in aldosterone secretion in response to angiotensin infusion (DAVIS et al., 1969).

5. Aldosterone secretion or excretion rate was found to be within the normal range in about 50% of patients with untreated congestive heart failure; most of these patients showed highly increased levels of arterial plasma angiotensin and peripheral blood renin activity (GENEST et al., 1968).

Moreover, decreases in aldosterone secretion have been observed in instances in which plasma renin activity remained normal:

1. Suppression of aldosterone secretion in man by treatment with glucocorticosteroids or ACTH was not consistently associated with decreases in plasma renin activity (NEWTON and LARAGH, 1968 a, b).

2. Medroxyprogesterone acetate partially suppressed aldosterone excretion in 5 normal subject—by a mechanism which was dependent on sodium intake—but did not decrease plasma renin activity (CRANE and HARRIS, 1969 a).

3. During adaptation to mild hypoxia, a slight increase in plasma renin activity was accompanied by a significant decrease in aldosterone secretion (SLATER et al., 1969).

In the light of all this evidence, I have serious doubts about a predominance of the renin-angiotensin system in the physiological control of aldosterone secretion in man or in the dog. There is even less evidence indicating that the renin-angiotensin system is of primary importance in the regulation of aldosterone production in the sheep and in the rat. If major alterations in the activity of the renin-angiotensin system are not regularly followed by homodirectional alterations in aldosterone secretion, the frequently observed parallelism of these two parameters may not be due to coincidence but it does not prove a simple direct causal relation either. At least, a causal relation appears to depend on other factors. Such factors may regulate the responsiveness of the adrenal cortex to angiotensin, for instance by activation or inactivation of the enzymes involved in the final steps of aldosterone biosynthesis.

8. Multifactorial Control of Aldosterone Biosynthesis

In the preceding chapter I have given my reasons for doubting that the renin-angiotensin system is the predominant factor in the physiological control of aldosterone secretion. I do not know if such a predominant factor really exists. In Fig. 19 I have listed all known and unknown factors influencing aldosterone biosynthesis of the rat adrenal cortex according to the evidence cited in this review without consideration of their relative physiological importance. In drawing this model, I have assumed that there is only one type of aldosterone-producing cells and that there is only one pathway of aldosterone biosynthesis. Even in this multifactorial model angiotensin II could be the most important factor, provided that the activity of all the other factors remained relatively constant. However, at least theoretically, the unknown factors acting on the final steps of aldosterone biosynthesis appear to be in a key position. It is of particular interest that potassium deficiency in some way leads to a decreased activity of the enzymes involved in the conversion of deoxycorticosterone to aldosterone, because potassium loss itself can be a consequence of elevated aldosterone secretion. This relationship could be the basis of a relatively simple feed-back control mechanism of aldosterone production. Thus, stimulation of aldosterone biosynthesis by an "adrenoglomerulotropin" such as angiotensin II would be self-limiting by the potassium loss it induces, unless this effect was offset by a factor stimulating the final steps of aldosterone biosynthesis.

In the past, the discovery of the functional interrelations between the zona glomerulosa of the adrenal cortex and the renin-angiotensin system has been the greatest achievement in the investigation of the physiological control of aldosterone secretion. I am convinced that the discovery of the nature and mechanism of action of the physiological factor or factors controlling the final steps of aldosterone biosynthesis could be an equally important achievement of future research.

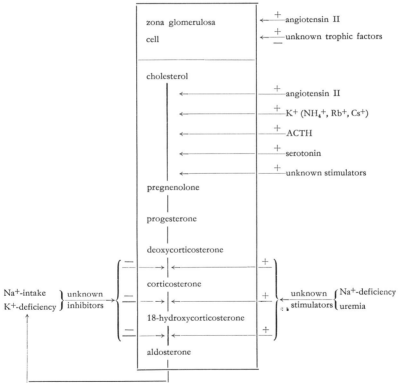

Fig. 19. Regulation of aldosterone biosynthesis in the rat

References

ABBOTT, E. C., GORNALL, A. G., SUTHERLAND, D. J. A., STIEFEL, M., LAIDLAW, J. C.: The influence of a heparin-like compound on hypertension, electrolytes and aldosterone in man. Canad. med. Ass. J. **94,** 1155 (1966b).

— MONKHOUSE, F. C., STEINER, J. W., LAIDLAW, J. C.: Effect of a sulfated mucopolysaccaride (RO 1—8307) on the zona glomerulosa of the rat adrenal gland. Endocrinology **78,** 651, (1966a).

AMES, R. P., BORKOWSKI, A. J., SICINSKI, A. M., LARAGH, J. H.: Prolonged infusions of angiotensin II and norepinephrine and blood pressure, electrolyte balance, and aldosterone and cortisol secretion in normal man and in cirrhosis with ascites. J. clin. Invest. **44,** 1171 (1965).

AYRES, P. J., EICHHORN, J., HECHTER, O., SABA, N., TAIT, J. F., TAIT, S. A. S.: Some studies on the biosynthesis of aldosterone and other adrenal steroids. Acta endocr. (Kbh.) **33,** 27 (1960).

— GOULD, R. P., SIMPSON, S. A., TAIT, J. F.: The *in vitro* demonstration of different corticosteroid production within the ox adrenal gland. Biochem. J. (abstract) **63,** 19 *P* (1956).

— HECHTER, O., SABA, N., SIMPSON, S. A., TAIT, J. F.: Intermediates in the biosynthesis of aldosterone by capsule strippings of ox adrenal gland. Biochem. J. (abstract) **65,** 22 *P* (1957).

BACH, I., BRAUN, S., GATI, T., SOS, J., UDVARDY, A.: Die Wirkung der einwertigen Kationen auf die Nebennierenrinde (abstract). Abstracts 1st int. Congr. Endocr. Copenhagen 1960, p. 151.

BALIKIAN, H. M., BRODIE, A. H., DALE, S. L., MELBY, J. C., TAIT, J. F.: Effect of posture on the metabolic clearance rate, plasma concentration and blood production rate of aldosterone in man. J. clin. Endocr. **28,** 1630 (1968).

BANIUKIEWICZ, S., BRODIE, A., FLOOD, C., MOTTA, M., OKAMOTA, M., TAIT, J. F., TAIT, S. A. S., BLAIR-WEST, J. R., COGHLAN, J. P., DENTON, D. A., GODING, J. R., SCOGGINS, B. A., WINTOUR, E. M., WRIGHT, R. D.: Adrenal biosynthesis of steroids *in vitro* and *in vivo* using continuous superfusion and infusion procedures. In: Functions of the Adrenal Cortex, Vol. 1. Ed.: K. W. McKERNS, Amsterdam: North Holland 1968, p. 153.

BARBOUR, B. H., SLATER, J. D. H., CASPER, A. G. T., BARTTER, F. C.: On the role of the central nervous system in control of aldosterone secretion. Life Sci. **4,** 1161 (1965).

BARRACLOUGH, M. A.: Sodium and water depletion with acute malignant hypertension. Amer. J. Med. **40,** 265 (1966).

BARRACLOUGH, M. A., BACCHUS, B., BROWN, J. J., DAVIES, D. L., LEVER, A. F., ROBERTSON, J. I. S.: Plasma-renin and aldosterone secretion in hypertensive patients with renal or renal arterial lesions. Lancet 1965 II, 1310.

BARTTER, F. C., BARBOUR, B. H., CARR, A. A., DELEA, C. S.: On the role of potassium and of the central nervous system in the regulation of aldosterone secretion. In: Aldosterone. Eds.: E. E. BAULIEU and P. ROBEL. Oxford: Blackwell 1964, p. 221.

— FOURMAN, P.: A non-renal effect of adrenal cortical steroids upon potassium metabolism. J. clin. Invest. (abstract) 36, 872 (1957).

BAULIEU, E. E., DE VIGAN, M., BRICAIRE, H., JAYLE, M. F.: Lack of plasma cortisol and urinary aldosterone in a pregnant woman with Addison's disease. J. clin. Endocr. 17, 1478 (1957).

BECK, J. C., McGARRY, E. E., DYRENFURTH, I., MORGEN, R. O., BIRD, E. D., VENNING, E. H.: Primate growth hormone studies in man. Metabolism 9, 699 (1960).

BIGLIERI, E. G., HANE, S., SLATON, P. E., JR., FORSHAM, P. H.: In vivo and in vitro studies of adrenal secretions in Cushing's syndrome and primary aldosteronism. J. clin. Invest. 42, 516 (1963).

— HERRON, M. A., BRUST, N.: 17-Hydroxylation deficiency in man. J. clin. Invest. 45, 1946 (1966b).

— SCHAMBELAN, M., SLATON, P. E., JR.: Effect of adrenocorticotropin on desoxycorticosterone, corticosterone and aldosterone excretion. J. clin. Endocr. 29, 1090 (1969).

— SLATON, P. E., JR.: Pregnancy and primary aldosteronism. J. clin. Endocr. 27, 1628 (1967).

— — KRONFIELD, S. J., SCHAMBELAN, M.: Diagnosis of an aldosterone-producing adenoma in primary aldosteronism. An evaluative maneuver. J. Amer. med. Ass. 201, 510 (1967).

— — SCHAMBELAN, M., KRONFIELD, S. J.: Hypermineralocorticoidism. Amer. J. Med. 45, 170 (1968).

— — SILEN, W. S., GALANTE, M., FORSHAM, P. H.: Postoperative studies of adrenal function in primary aldosteronism. J. clin. Endocr. 26, 553 (1966a).

— WATLINGTON, C. O., FORSHAM, P. H.: Sodium retention with human growth hormone and its subfractions. J. clin. Endocr. 21, 361 (1961).

BINNION, P. F., DAVIS, J. O., BROWN, T. C., OLICHNEY, M. J.: Mechanisms regulating aldosterone secretion during sodium depletion. Amer. J. Physiol. 208, 655 (1965).

BIRMINGHAM, M. K., ELLIOTT, F. H., VALERE, P. H. L.: The need for the presence of calcium for the stimulation in vitro of rat adrenal glands by adrenocorticotrophic hormone. Endocrinology 53, 687 (1953).

BIRON, P., CHRETIEN, M., KOIW, E., GENEST, J.: Effects of angiotensin infusions on aldosterone and electrolyte excretion in normal subjects and patients with hypertension and adrenocortical disorders. Brit. med. J. 1962 I, 1569.

BIRON, P., KOIW, E., NOWACZYNSKI, W., BROUILLET, J., GENEST, J.: The effects of intravenous infusions of valine-5-angiotensin II and other pressor agents on urinary electrolytes and corticosteroids, including aldosterone. J. clin. Invest. **40,** 338 (1961).

BLACK, D. A. K., MILNE, M. D.: Experimental potassium depletion in man. Clin. Sci. **11,** 397 (1952).

BLAIR-WEST, J. R., BRODIE, A., COGHLAN, J. P., DENTON, D. A., FLOOD, C., GODING, J. R., SCOGGINS, B. A., TAIT, J. F., TAIT, S. A. S., WINTOUR, E. M., WRIGHT, R. D.: Studies on the biosynthesis of aldosterone using sheep adrenal transplant: Effect of sodium depletion on the conversion of corticosterone to aldosterone. J. Endocr. **46,** 453 (1970b).

— CAIN, M. D., CATT, K. J., COGHLAN, J. P., DENTON, D. A., FUNDER, J. W., HARDING, R., McKENZIE, J. S., SCOGGINS, B. A., WINTOUR, M., WRIGHT, R. D.: Does local sodium concentration specifically determine angiotensin action? Acta endocr. (Kbh.) (abstract) Suppl. **138,** 124 (1969a).

— — — — — — SCOGGINS, B. A., WINTOUR, M., WRIGHT, R. D.: The control of aldosterone secretion in sodium deficiency. Acta endocr. (Kbh.) (abstract) Suppl. **138,** 125 (1969b).

— COGHLAN, J. P., DENTON, D. A.: Evidence against an aldosterone feed-back mechanism within the adrenal gland. Acta endocr. (Kbh.) **41,** 61 (1962).

— — — GODING, J. R., MUNRO, J. A., PETERSON, R. E., WINTOUR, M.: Humoral stimulation of adrenocortical secretion. J. clin. Invest. **41,** 1606 (1962).

— — — — — WRIGHT, R. D.: The effect of neuraxial ablations upon the secretion of electrolyte-active steroid. J. Physiol. (Lond.) **153,** 53 P (1960).

— — — — ORCHARD, E., SCOGGINS, B., WINTOUR, M., WRIGHT, R. D.: Mechanisms regulating aldosterone secretion during sodium deficiency. In: Proc. 3rd int. Congr. Nephrol., Washington 1966, Vol. 1. Basel-New York: Karger 1967, p. 201.

— — — — WINTOUR, M., WRIGHT, R. D.: The control of aldosterone secretion. Recent Progr. Hormone Res. **19,** 311 (1963).

— — — — — — Effect of variations of plasma sodium concentration on the adrenal response to angiotensin II. Circulation Res. **17,** 386 (1965a).

— — — — — — A study of feedback in aldosterone secretion. Endocrinology **77,** 501 (1965b).

— — — — — — The direct effects of increased sodium concentration in adrenal arterial blood on corticosteroid secretion in sodium deficient sheep. Aust. J. exp. Biol. med. Sci. **44,** 455 (1966).

— — — — — — The effect of nephrectomy on aldosterone secretion in the conscious sodium-depleted hypophysectomized sheep. Aust. J. exp. Biol. med. Sci. **46,** 295 (1968b).

BLAIR-WEST, J. R., COGHLAN, J. P., DENTON, D. A., GODING, J. R., WINTOUR, M., WRIGHT, R. D.: The local action of ammonium, calcium and magnesium on adrenocortical secretion. Aust. J. exp. Biol. med. Sci. **46,** 371 (1968 d).

— — — MUNRO, J. A., WINTOUR, M., WRIGHT, R. D.: The effect of bilateral nephrectomy and midcollicular decerebration with pinealectomy and hypophysectomy on the corticosteroid secretion of sodium-deficient sheep. J. clin. Invest. **43,** 1576 (1964).

— — — NELSON, J. F., ORCHARD, E., SCOGGINS, B. A., WRIGHT, R. D., MYERS, K., JUNQUEIRA, C. L.: Physiological, morphological and behavioural adaptation to a sodium deficient environment by wild native Australian and introduced species of animals. Nature (Lond.) **217,** 922 (1968a).

— — — ORCHARD, E., SCOGGINS, B. A., WRIGHT, R. D.: Renin-angiotensin-aldosterone system and sodium balance in experimental renal hypertension. Endocrinology **83,** 1199 (1968c).

— — — SCOGGINS, B. A., WINTOUR, E. M., WRIGHT, R. D.: The onset of effects of ACTH, angiotensin II and raised plasma potassium concentration on the adrenal cortex. Steroids **15,** 433 (1970a).

BLEDSOE, T., ISLAND, D. P., LIDDLE, G. W.: Studies on the mechanism through which sodium depletion increases aldosterone biosynthesis in man. J. clin. Invest. **45,** 524 (1966).

— — RIONDEL, A. M., LIDDLE, G. W.: Modification of aldosterone secretion and electrolyte excretion in man by a chemical inhibitor of 18-oxidation. J. clin. Endocr. **24,** 740 (1964).

BOJESEN, E.: Concentrations of aldosterone and corticosterone in peripheral plasma of rats. The effects of salt depletion, potassium loading and intravenous injections of renin and angiotensin II. Eur. J. Steroids **1,** 145 (1966).

— DEGN, H.: A double isotope derivative method for the determination of aldosterone in plasma. The effect of adrenalectomy on aldosterone level in anaesthethesized dogs. Acta endocr. (Kbh.) **37,** 541 (1961).

BOYD, G. W., ADAMSON, A. R., JAMES, V. H. T., PEART, W. S.: The role of the renin-angiotensin system in the control of aldosterone in man. Proc. roy. Soc. Med. **62,** 1253 (1969).

BOYD, J., MARUSIC, E., MULROW, P.: Effect of potassium on aldosterone biosynthesis by rat adrenal mitochondria. Clin. Res. (abstract) **16,** 262 (1968).

BRODIE, A. H., SHIMIZU, N., TAIT, S. A. S., TAIT, J. F.: A method for the measurement of aldosterone in peripheral plasma using ^{3}H-acetic anhydride. J. clin. Endocr. **27,** 997 (1967).

BROWN, J. J., CHINN, R. H., DAVIES, D. L., DÜSTERDIECK, G., FRASER, R., LEVER, A. F., ROBERTSON, J. I. S., TREE, M., WISEMAN, A.: Plasma electrolytes, renin, and aldosterone in the diagnosis of primary aldosteronism. With a note on plasma-corticosterone concentration. Lancet **1968 II,** 55.

Brown, J. J., Davies, D. L., Doak, P. B., Lever, A. F., Robertson, J. I. S.: Plasma renin in normal pregnancy. Lancet 1963 II, 900.

— — — — — Plasma renin concentration in the hypertensive diseases of pregnancy. J. Obstet. Gynaec. Brit. Cwlth 73, 410 (1966 b).

— — — — — Tree, M.: The presence of renin in the amniotic fluid. Lancet 1964 II c, 64.

— — — — — Trust, P.: Plasma-renin concentration in hypertensive disease of pregnancy. Lancet 1965 II b, 1219.

— — Lever, A. F., McPherson, D., Robertson, J. I. S.: Plasma renin concentration in relation to changes in posture. Clin. Sci. 30, 279 (1966 c).

— — — Robertson, J. I. S.: Influence of sodium deprivation and loading on the plasma-renin in man. J. Physiol. (Lond.) 173, 408 (1964 a).

— — — — Variation in plasma renin during the menstrual cycle. Brit. med. J. 1964 II b, 1114.

— — — — Plasma renin concentration in human hypertension. 1: Relationship between Renin, Sodium, and Potassium. Brit. med. J. 1965 II a, 144.

— — — — Renin and angiotensin. A survey of some aspects. Postgrad. med. J. 42, 153 (1966 a).

Brownie, A. C., Skelton, F. R.: The metabolism of progesterone-4-^{14}C by adrenal homogenates from rats with adrenal-regeneration hypertension. Steroids 6, 47 (1965).

Bryan, G. T., Kliman, B., Gill, J. R., Jr., Bartter, F. C.: Effect of human renin on aldosterone secretion rate in normal man and in patients with the syndrome of hyperaldosteronism, juxtaglomerular hyperplasia and normal blood pressure. J. clin. Endocr. 24, 729 (1964).

Bunge, G.: Über die Bedeutung des Kochsalzes und das Verhalten der Kalisalze im menschlichen Organismus. Z. Biol. 9, 104 (1873).

Burrow, G. N.: Steroid inhibition of aldosterone biosynthesis in adrenal mitochondria. Clin. Res. (abstract) 16, 338 (1968).

— Mulrow, P. J., Bondy, P. K.: Protein synthesis and aldosterone production. Endocrinology 79, 955 (1966).

Burwell, L. R., Davis, W. W., Bartter, F. C.: Studies on the loci of action of stimuli to the biogenesis of aldosterone. Proc. roy. Soc. Med. 62, 1254 (1969).

Bush, I. E.: Species differences in adrenocortical secretion. J. Endocr. 9, 95 (1953).

Cade, R., Perenich, T.: Secretion of aldosterone by rats. Amer. J. Physiol. 208, 1026 (1965).

Cannon, P. J., Ames, R. P., Laragh, J. H.: Relation between potassium balance and aldosterone secretion in normal subjects and in patients with hypertensive or renal tubular disease. J. clin. Invest. 45, 865 (1966).

Cantin, M., Veilleux, R.: The juxtaglomerular apparatus and zona glomerulosa in magnesium-deficient rats. Rev. Canad. Biol. 27, 179 (1968).

CAPELLI, J. P., WESSON, L. G., JR., APONTE, G. E., FARALDO, C., JAFFE, E.: Characterization and source of a renin-like enzyme in anephric humans. J. clin. Endocr. **28,** 221 (1968).

CARPENTER, C. C. J., DAVIS, J. O., AYERS, C. R.: Relation of renin, angiotensin II, and experimental renal hypertension to aldosterone secretion. J. clin. Invest. **40,** 2026 (1961).

CARRETERO, O., GROSS, F.: Renin substrate in plasma under various experimental conditions in the rat. Amer. J. Physiol. **213,** 695 (1967).

CARSTENSEN, H., BURGERS, A. C. J., LI, C. H.: Demonstration of aldosterone and corticosterone as the principal steroids formed in incubates of adrenals of the American bullfrog *(Rana catesbeiana)* and stimulation of their production by mammalian adrenocorticotropin. Gen. comp. Endocr. **1,** 37 (1961).

CEJKA, V., DE VRIES, L. A., SMORENBERG-SCHOORL, M. E., VAN DAATSELAAR, J. J., BORST, J. G. G., MAJOOR, C. L. H.: Effect of heparinoid and spirolactone on the renal excretion of sodium and aldosterone. Lancet **1960I,** 317.

CHAGOYA, L., NURKO, B., SANTOS, E., RIVERA, A.: 6-chloro-, 6-dehydro, 17α-acetoxyprogesterone: its possible action as an aldosterone antagonist. J. clin. Endocr. **21,** 1364 (1961).

CHANNICK, B. J., ADLIN, E. V., MARKS, A. D.: Suppressed plasma renin activity in hypertension. Arch. intern. Med. **123,** 131 (1969).

COGHLAN, J. P., BLAIR-WEST, J. R.: Aldosterone. In: Hormones in Blood, Vol. 2. 2nd ed. Eds.: C. H. GRAY and A. L. BACHARACH. New York: Academic Press 1967, p. 391.

— SCOGGINS, B. A.: Measurement of aldosterone in peripheral blood of man and sheep. J. clin. Endocr. **27,** 1470 (1967a).

— — The measurement of aldosterone, cortisol and corticosterone in the blood of the wombat *(Vombatus Hirsutus* Perry) and the kangaroo *(Macropus Giganteus).* J. Endocr. **39,** 445 (1967b).

COHEN, M. P.: Aminoglutethimide inhibition of adrenal desmolase activity. Proc. Soc. exp. Biol. (N. Y.) **127,** 1086 (1968).

COHEN, R. B.: Effects of long-term sodium deprivation on the adrenal cortices of rats: a histochemical study. Endocrinology **77,** 1043 (1965).

CONN, J. W., COHEN, E. L., ROVNER, D. R.: Suppression of plasma renin activity in primary aldosteronism. Distinguishing primary from secondary aldosteronism in hypertensive disease. J. Amer. med. Ass. **190,** 213 (1964).

— ROVNER, D. R., COHEN, E. L.: Licorice-induced pseudoaldosteronism. Hypertension, hypokalemia, aldosteronopenia, and suppressed plasma renin activity. J. Amer. med. Ass. **205,** 492 (1968).

— — ANDERSON, J. E., JR.: Inhibition by heparinoid of aldosterone biosynthesis in man. J. clin. Endocr. **26,** 527 (1966).

CONNORS, M., ROSENKRANTZ, H.: Serotonin uptake and action on the adrenal cortex. Endocrinoloy **71,** 407 (1962).

COPE, C. L., HARWOOD, M., PEARSON, J.: Aldosterone secretion in hypertensive diseases. Brit. med. J. **1962I**, 659.

COPPAGE, W. S., JR., ISLAND, D., SMITH, M., LIDDLE, G. W.: Inhibition of aldosterone secretion and modification of electrolyte excretion by a chemical inhibitor of 11β-hydroxylation. J. clin. Invest. **38**, 2101 (1959).

CORTES, J. M., PERON, F. G., DORFMAN, R. I.: Secretion of 18-hydroxydeoxycorticosterone by the rat adrenal gland. Endocrinology **73**, 713 (1963).

CRABBE, J.: The Sodium-Retaining Action of Aldosterone. Bruxelles: Editions Arscia S.A. & Presses Académiques Européennes 1963.

— REDDY, W. J., ROSS, E. J., THORN, G. W.: The stimulation of aldosterone secretion by adrenocorticotropic hormone (ACTH). J. clin. Endocr. **19**, 1185 (1959).

CRANE, M. G., HARRIS, J. J.: Plasma renin activity and aldosterone excretion rate in normal subjects. I. Effect of ethinyl estradiol and medroxyprogesterone acetate. J. clin. Endocr. **29**, 550 (1969a).

— — Plasma renin activity and aldosterone excretion rate in normal subjects. II. Effect of oral contraceptives. J. clin. Endocr. **29**, 558 (1969b).

— — Desoxycorticosterone secretion rate studies in edematous patients. Amer. J. med. Sci. **259**, 27 (1970).

CSANKY, M. F. D., VAN DER WAL, B., DE WIED, D.: The regulation of aldosterone production in normal and sodium-deficient rats. J. Endocr. **41**, 179 (1968).

CUSHMAN, P., JR.: Inhibition of aldosterone secretion by ouabain in dog adrenal tissue. Endocrinology **84**, 808 (1969).

— ALTER, S., HILTON, J. G.: Effects of various growth hormone preparations on dog adrenal cortical secretion. Endocrinology **78**, 971 (1966).

DAVID, R., GOLAN, S., DRUCKER, W.: Familial aldosterone deficiency: enzyme defect, diagnosis and clinical course. Pediatrics **41**, 403 (1968).

DAVIS, J. O.: Mechanisms regulating the secretion and metabolism of aldosterone in experimental secondary hyperaldosteronism. Recent Progr. Hormone Res. **17**, 293 (1961).

— The control of aldosterone secretion. Physiologist **5**, 65 (1962).

— The regulation of aldosterone secretion. In: The Adrenal Cortex. Ed.: A. B. EISENSTEIN. Boston: Little, Brown & Co. 1967, p. 203.

— ANDERSON, E., CARPENTER, C. C. J., AYERS, C. R., HAYMAKER, W., SPENCE, W. T.: Aldosterone and corticosterone secretion following midbrain transection. Amer. J. Physiol. **200**, 437 (1961c).

— AYERS, C. R., CARPENTER, C. J.: Renal origin of an aldosterone-stimulating hormone in dogs with thoracic caval constriction and in sodium-depleted dogs. J. clin. Invest. **40**, 1466 (1961b).

— BAHN, R. C., BALL, W. C., JR.: Subacute and chronic effects of hypothalamic lesions on aldosterone and sodium excretion. Amer. J. Physiol. **197**, 387 (1959b).

DAVIS, J. O., BAHN, R. C., YANKOPOULOS, N. A., KLIMAN, B., PETERSON, R. E.: Acute effects of hypophysectomy and diencephalic lesions on aldosterone secretion. Amer. J. Physiol. **197,** 380 (1959a).
— BINNION, P. F., BROWN, T. C., JOHNSTON, C. I.: Mechanisms involved in the hypersecretion of aldosterone during sodium depletion. Circulation Res. **18/19,** Suppl. 1, 1—143 (1966).
— CARPENTER, C. C. J., AYERS, C. R., BAHN, R. C.: Relation of anterior pituitary function to aldosterone and corticosterone secretion in conscious dogs. Amer. J. Physiol. **199,** 212 (1960b).
— — — HOLMAN, J. E., BAHN, R. C.: Evidence for secretion of an aldosterone-stimulating hormone by the kidney. J. clin. Invest. **40,** 684 (1961a).
— HOWARDS, S. S., JOHNSTON, C. I., WRIGHT, F. S.: Deoxycorticosterone secretion in chronic experimental heart failure and during infusions of angiotensin II. Proc. Soc. exp. Med. (N.Y.) **127,** 164 (1968).
— URQUHART, J., HIGGINS, J. T., JR.: The effects of alterations of plasma sodium and potassium concentration on aldosterone secretion. J. clin. Invest. **42,** 597 (1963).
— — — RUBIN, E. C., HARTROFT, P. M.: Hypersecretion of aldosterone in dogs with a chronic aortic-caval fistula and high output heart failure. Circulation Res. **14,** 471 (1964).
— YANKOPOULOS, N. A., LIEBERMAN, F., HOLMAN, J., BAHN, R. C.: The role of the anterior pituitary in the control of aldosterone secretion in experimental secondary aldosteronism. J. clin. Invest. **39,** 765 (1960a).
DAVIS, W. W., BURWELL, L. R., BARTTER, F. C., Inhibition of the effects of angiotensin II on adrenal steroid production by dietary sodium. Proc. nat. Acad. Sci. **63,** 718 (1969).
— — CASPER, A. G. T., BARTTER, F. C.: Sites of action of sodium depletion on aldosterone biosynthesis in the dog. J. clin. Invest. **47,** 1425 (1968).
— GARREN, L. D.: On the mechanism of action of adrenocorticotropic hormone. The inhibitory site of cycloheximide in the pathway of steroid biosynthesis. J. biol. Chem. **243,** 5153 (1968).
— NEWSOME, H. N., JR., WRIGHT, L. D., JR., HAMMOND, W. G., EASTON, J., BARTTER, F. C.: Bilateral adrenal hyperplasia as a cause of primary aldosteronism with hypertension, hypokalemia and suppressed renin activity. Amer. J. Med. 42, 642 (1967).
DEANE, H. W., SHAW, J. H., GREEP, R. O.: The effect of altered sodium or potassium intake on the width and cytochemistry of the zona glomerulosa of the rat's adrenal cortex. Endocrinology **43,** 133 (1948).
DEGENHART, H. J., FRANKENA, L., VISSER, H. K. A., COST, W. S., VAN SETERS, A. P.: Further investigation of a new hereditary defect in the biosynthesis of aldosterone: evidence for a defect in 18-hydroxylation of corticosterone. Acta physiol. pharmacol. Neerl. **14,** 1 (1966).
DENTON, D. A., GODING, J. R., WRIGHT, R. D.: Control of adrenal secretion of electrolyte-active steroids. Brit. med. J. **1959II,** 447 and 522.

DEXTER, R. N., FISHMAN, L. M., NEY, R. L., LIDDLE, G. W.: Inhibition of adrenal corticosteroid synthesis by aminoglutethimide: Studies on the mechanism of action. J. clin. Endocr. **27,** 473 (1967).

DOLLERY, C. T., SHACKMAN, R., SHILLINGFORD, J.: Malignant hypertension and hypokalemia: cured by nephrectomy. Brit. med. J. **1959 II,** 1367.

DONALDSON, E. M., HOLMES, W. N.: Corticosteroidogenesis in the fresh water- and saline-maintained duck *(Anas platyrhynchos).* J. Endocr. **32,** 329 (1965).

— — STACHENKO, J.: *In vitro* corticosteroidogenesis by the duck *(Anas playtyrhynchos)* adrenal. Gen. comp. Endocr. **5,** 542 (1965).

DRIESSEN, W. M. M., BENRAAD, T. J., KLOPPENBORG, P. W. C.: Effect of potassium loading on aldosterone secretory rate and plasma renin activity in normals and hypertensives. Acta endocr. (Kbh.) (abstract) Suppl. **138,** 126 (1969).

DUFAU, M. L., KLIMAN, B.: Pharmacologic effects of angiotensin-II-amide on aldosterone and corticosterone secretion by the intact anesthetized rat. Endocrinology **82,** 29 (1968a).

— — Acute effects of angiotensin-II-amide on aldosterone and corticosterone secretion by morphine-pentobarbital treated rats. Endocrinology **83,** 180 (1968b).

DYRENFURTH, I., LUCIS, O. J., BECK, J. C., VENNING, E. H.: Studies in patients with adrenocortical hyperfunction. III. *In vitro* secretion of steroids by human adrenal glands. J. clin. Endocr. **20,** 765 (1960).

EBERLEIN, W. R., BONGIOVANNI, A. M.: Congenital adrenal hyperplasia with hypertension: unusual steroid pattern in blood and urine. J. clin. Endocr. **15,** 1531 (1955).

EHRLICH, E. N., DOMINGUEZ, O. V., SAMUELS, L. T., LYNCH, D., OBERHELMAN, H., JR., WARNER, N. E.: Aldosteronism and precocious puberty due to an ovarian androblastoma (Sertoli cell tumor). J. clin. Endocr. **23,** 358 (1963).

— LAVES, M., LUGIBIHL, K., LANDAU, R. L.: Progesterone-aldosterone interrelationsships in pregnancy. J. Lab. clin. Med. **59,** 588 (1962).

EILERS, E. A., PETERSON, R. E.: Aldosterone secretion in the rat. In: Aldosterone. Eds.: E. E. BAULIEU and P. ROBEL. Oxford: Blackwell 1964, p. 251.

EISENSTEIN, A. B., HARTROFT, P. M.: Alterations in the rat adrenal cortex induced by sodium deficiency: steroid hormone secretion. Endocrinology **60,** 634 (1957).

ELEMA, J. D., HARDONK, M. J., KOUDSTAAL, J., ARENDS, A.: Acute enzyme histochemical changes in the zona glomerulosa of the rat adrenal cortex: I. The effect of peritoneal dialysis with a glucose 5% solution. Acta endocr. (Kbh.) **59,** 508 (1968a).

— — — — Acute enzyme histochemical changes in the zona glomerulosa of the rat adrenal cortex: II. The effect of bilateral nephrectomy either alone or followed by peritoneal dialysis with 5% glucose. Acta endocr. (Kbh.) **59**, 519 (1968b).

FABRE, L. F., JR., BANKS, R. C., McISAAC, W. M., FARRELL, G.: Effects of ubiquinone and related substances on secretion of aldosterone and cortisol. Amer. J. Physiol. **208**, 1275 (1965).

FALBRIARD, A., MULLER, A. F., NEHER, R., MACH, R. S.: Etude des variations de l'aldostéronurie sous l'effet de surcharges en potassium et de déperditions rénales et extrarénales de sel et d'eau. Schweiz. med. Wschr. **85**, 1218 (1955).

FARESE, R. V.: Effects of Actinomycin D on ACTH-induced corticosteroidogenesis. Endocrinology **78**, 929 (1966).

— Regulation of adrenal growth and steroidogenesis by ACTH. In: Functions of the Adrenal Cortex, Vol. 1. Ed.: K. W. McKERNS. Amsterdam: North Holland 1968, p. 539.

FARRELL, G.: Adrenoglomerulotropin. Circulation **21**, 1009 (1960).

— McISAAC, W. M.: Adrenoglomerulotropin. Arch. Biochem. **94**, 543 (1961).

— RAUSCHKOLB, E. W., ROYCE, P. C.: Secretion of aldosterone by the adrenal of the dog: Effects of hypophysectomy and ACTH. Amer. J. Physiol. **182**, 269 (1955).

FAZEKAS, A. G., KOKAI, K.: Biosynthesis of C-18-oxygenated corticosteroids by rabbit adrenals *in vitro*. Steroids **9**, 177 (1967).

FEKETE, G., SZEBERENYI, S.: Comparison of the aldosterone production suppressing activity of synthetic glucocorticoids. IV. Conf. Hung. Ther. Invest. Pharmacol. 373 (1968).

FERGUSON, J. J., JR.: Metabolic Inhibitors and Adrenal Function. In: Functions of the Adrenal Cortex, Vol. 1. Ed.: K. W. McKERNS. Amsterdam: North Holland 1968, p. 463.

— MORITA, Y.: RNA synthesis and adrenocorticotropin responsiveness. Biochim. biophys. Acta (Amst.) **87**, 348 (1964).

FINKELSTEIN, J. W., KOWARSKI, A., SPAULDING, J. S., MIGEON, C. J.: Effect of various preparations of human growth hormone on aldosterone secretion rate of hypopituitary dwarfs. Amer. J. Med. **38**, 517 (1965).

FISHMAN, L. M., KÜCHEL, O., LIDDLE, G. W., MICHELAKIS, A. M., GORDON, R. D., CHICK, W. T.: Incidence of primary aldosteronism in uncomplicated „essential hypertension". A prospective study with elevated aldosterone secretion and suppressed plasma renin activity used as diagnostic criteria. J. Amer. med. Ass. **205**, 497 (1968).

— LIDDLE, G. W., ISLAND, D. P., FLEISCHER, N., KÜCHEL, O.: Effects of amino-glutethimide on adrenal function in man. J. clin. Endocr. **27**, 481 (1967).

FORD, H. C., BAILEY, R. E.: The effect of heparin on aldosterone secretion and metabolism in primary aldosteronism. Steroids **7**, 30 (1966).

FRAGACHAN, F., NOWACZYNSKI, W., BERTRANEAU, E., KALINA, M., GENEST, J.: Evidence of *in vivo* inhibition of 11β-hydroxylation of steroids by dehydroepiandrosterone in the dog. Endocrinology **84,** 98 (1969).

FUNDER, J. W., BLAIR-WEST, J. R., COGHLAN, J. P., DENTON, D. A., SCOGGINS, B. A., WRIGHT, R. D.: Effect of plasma [K+] on the secretion of aldosterone. Endocrinology **85,** 381 (1969).

GANN, D. S., CRUZ, J. F., CASPER, A. G. T., BARTTER, F. C.: Mechanism by which potassium increases aldosterone secretion in the dog. Amer. J. Physiol. **202,** 991 (1962).

— DELEA, C. S., GILL, J. R., JR., THOMAS, J. P., BARTTER, F. C.: Control of aldosterone secretion by change of body potassium in normal man. Amer. J. Physiol. **207,** 104 (1964).

— TRAVIS, R. H.- Mechanisms of hemodynamic control of secretion of aldosterone in the dog. Amer. J. Physiol. **207,** 1095 (1964).

GANONG, W. F., BIGLIERI, E. G., MULROW, P. J.: Mechanisms regulating adrenocortical secretion of aldosterone and glucocorticoids. Recent Progr. Hormone Res. **22,** 381 (1966).

— BORYCZKA, A. T.: Effect of a low sodium diet on the aldosterone-stimulating activity of angiotensin II in dogs. Proc. Soc. exp. Biol. (N.Y.) **124,** 1230 (1967).

— — SHACKLEFORD, R.: Effect of renin on adrenocortical sensitivity to ACTH and angiotensin II in dogs. Endocrinology **80,** 703 (1967a).

— — — CLARK, R. M., CONVERSE, R. P.: Effect of dietary sodium restriction on adrenal cortical response to ACTH. Proc. Soc. exp. Biol. (N.Y.) **118,** 792 (1965a).

— LEE, T. C., VAN BRUNT, E. E., BIGLIERI, E. G.: Aldosterone secretion in dogs immunized with hog renin. Endocrinology **76,** 1141 (1965b).

— LIEBERMAN, A. H., DAILY, W. J. R., YUEN, V. S., MULROW, P. J., LUETSCHER, J. A., JR., BAILEY, R. E.: Aldosterone secretion in dogs with hypothalamic lesions. Endocrinology **65,** 18 (1959).

— MULROW, P. J.: Evidence of secretion of an aldosterone-stimulating substance by the kidney. Nature (Lond.) **190,** 1115 (1961).

— — BORYCZKA, A., CERA, G.: Evidence for a direct effect of angiotensin-II on adrenal cortex of the dog. Proc. Soc. exp. Biol. (N.Y.) **109,** 381 (1962).

— PEMBERTON, D. L., VAN BRUNT, E. E.: Adrenocortical responsiveness to ACTH and angiotensin II in hypophysectomized dogs and dogs treated with large doses of glucocorticoids. Endocrinology **81,** 1147 (1967b).

— VAN BRUNT, E. E., LEE, T. C., MULROW, P. J.: Inhibition of aldosterone-stimulating activity of hog and dog renin by the plasma of dogs immunized with hog renin. Proc. Soc. exp. Biol. (N.Y.) **112,** 1062 (1963).

GAUNT, R., CHART, J. J., RENZI, A. A.: Inhibitors of adrenal cortical function. Ergebn. Physiol. **56,** 113 (1965).

— STEINETZ, B. G., CHART, J. J.: Pharmacological alteration of steroid hormone functions. Clin. Pharmacol. Ther. **9,** 657 (1968).

GEELHOED, G. W., VANDER, A. J.: The role of aldosterone in renin secretion. Life Sci. **6,** 525 (1967).

GENEST, J.: Angiotensin, aldosterone and human arterial hypertension. Canad. med. Ass. J. **84,** 403 (1961).

— BOUCHER, R., NOWACZYNSKI, W., KOIW, E., DE CHAMPLAIN, J., BIRON, P., CHRETIEN, M., MARC-AURELE, J.: Studies on the relationship of aldosterone and angiotensin to human hypertensive disease. In: Aldosterone, Eds.: E. E. BAULIEU and P. ROBEL. Oxford: Blackwell 1964, p. 393.

— GRANGER, P., DE CHAMPLAIN, J., BOUCHER, R.: Endocrine factors in congestive heart failure. Amer. J. Cardiol. **22,** 35 (1968).

GEORGE, J. M., GILLESPIE, L., BARTTER, F. C.: Aldosterone secretion in hypertension. Ann. intern. Med. **69,** 693 (1968).

GINN, H. E., CADE, R., McCALLUMM T., FREGLEY, M.: Aldosterone secretion in magnesium-deficient rats. Endocrinology **80,** 969 (1967).

GIROUD, C. J. P., SAFFRAN, M., SCHALLY, A. V., STACHENKO, J., VENNING, E. H.: Production of aldosterone by rat adrenal glands *in vitro.* Proc. Soc. exp. Biol. (N.Y.) **92,** 855 (1956b).

— STACHENKO, J., PILETTA, P.: *In vitro* studies of the functional zonation of the adrenal cortex and of the production of aldosterone. In: Aldosterone. Eds.: A. F. MULLER and C. M. O'CONNOR. Boston: Little, Brown & Co. 1958, p. 56.

— — VENNING, E. H.: Secretion of aldosterone by the zona glomerulosa of rat adrenal glands incubated *in vitro.* Proc. Soc. exp. Biol. (N.Y.) **92,** 154 (1956a).

GLAZ, E., SUGAR, K.: The effect of synthetic angiotensin II on synthesis of aldosterone by the adrenals. J. Endocr. **24,** 299 (1962).

— — Effect of heparin and heparinoids on the synthesis of aldosterone and corticosterone by the rat adrenal gland. Endocrinology **74,** 159 (1964).

GOLDSMITH, O., SOLOMON, D. H., HORTON, R.: Hypogonadism and mineralocorticoid excess. The 17-hydroxylase deficiency syndrome. New Engl. J. Med. **277,** 673 (1967).

GORDON, R. D., FISHMAN, L. M., LIDDLE, G. W.: Plasma renin activity and aldosterone secretion in a pregnant woman with primary aldosteronism. J. clin. Endocr. **27,** 385 (1967).

— WOLFE, L. K., ISLAND, D. P., LIDDLE, G. W.: A diurnal rhythm in plasma renin activity in man. J. clin. Invest. **45,** 1587 (1966).

GOULD, A. B., SKEGGS, L. T., KAHN, J. R.: Measurement of renin and substrate concentrations in human serum. Lab. Invest. **15,** 1802 (1966).

GOWENLOCK, A. H., WRONG, O.: Hyperaldosteronism secondary to renal ischemia. Quart. J. Med. **31,** 323 (1962).

GRAHAME-SMITH, D. G., BUTCHER, R. W., NEY, R. L., SUTHERLAND, E. W.: Adenosine 3',5'-monophosphate as the intracellular mediator of the action of adrenocorticotropic hormone on the adrenal cortex. J. biol. Chem. **242,** 5535 (1967).

GRAY, M. J., STRAUSFELD, K. S., WATANABE, M., SIMS, E. A. H., SOLO-
MON, S.: Aldosterone secretory rates in the normal menstrual cycle.
J. clin. Endocr. **28,** 1269 (1968).

GREENE, J. A., JR., VANDER, A. J., KOWALCZYK, R. S.: Plasma renin
activity and aldosterone excretion after renal homotransplantation.
J. Lab. clin. Med. **71,** 586 (1968).

GREENGARD, P., PSYCHOYOS, S., TALLAN, H. H., COOPER, D. Y., ROSEN-
THAL, O., ESTABROOK, R. W.: Aldosterone synthesis by adrenal mitochondria. III. Participation of cytochrome P-450. Arch. Biochem. **121,**
298 (1967).

GREENWAY, C. V., VERNEY, E. B.: The effect of adrenocorticotrophic
hormone on the secretion of corticosteroids by the isolated perfused
adrenal gland of the dog. J. Physiol. (Lond.) **162,** 183 (1962).

GREEP, R. O., DEANE, H. W.: Cytochemical evidence for the cessation of
hormone production in the zona glomerulosa of the rat's adrenal
cortex after prolonged treatment with desoxycorticosterone acetate.
Endocrinology **40,** 417 (1947).

GRÖGER-KORBER, E. M.: Beeinflussung der Nebennierenrindensekretion
durch Polypeptide *in vitro.* Arch. int. Pharmacodyn. **163,** 158 (1966).

GROSS, F.: Renin und Hypertensin, physiologische oder pathologische
Wirkstoffe? Klin. Wschr. **36,** 693 (1958).

— Differentiation of effects mediated by aldosterone and renin-angiotensin in rats with experimental hypertension. In: Aldosterone. Eds.:
E. E. BAULIEU and P. ROBEL. Oxford: Blackwell 1964, p. 307.

— The regulation of aldosterone secretion by the renin-angiotensin system
under various conditions. Acta endocr. (Kbh.) Suppl. **124,** 41 (1967).

— BRUNNER, H., ZIEGLER, M.: Renin-angiotensin system, aldosterone,
and sodium balance. Recent Progr. Hormone Res. **21,** 119 (1965).

— CARRETERO, O., JOHNSTON, C. I.: Effect of ureteral ligation on blood
pressure and renin content of the kidneys in renal hypertensive rats.
Proc. Soc. exp. Biol. (N.Y.) **128,** 935 (1968).

— SCHAECHTELIN, G., ZIEGLER, M., BERGER, M.: A renin-like substance
in the placenta and uterus of the rabbit. Lancet **1964I,** 914.

HALKERSTON, I. D. K.: Heterogeneity of the response of adrenal cortex
tissue slices to adrenocorticotropin. In: Functions of the Adrenal
Cortex, Vol. 1. Ed.: K. W. MCKERNS. Amsterdam: North Holland
1968, p. 399.

— FEINSTEIN, M., HECHTER, O.: Further observation on the inhibition
of adrenal protein synthesis by ACTH *in vitro.* Endocrinology **76,** 801
(1965).

— — — Effect of lytic enzymes upon the responsivity of rat adrenals *in
vitro.* I. Effect of trypsin upon the steroidogenic action of reduced
triphosphopyridine nucleotide. Endocrinology **83,** 61 (1968).

HARTROFT, P. M., EISENSTEIN, A. B.: Alterations in the adrenal cortex of
the rat induced by sodium deficiency: correlation of histological
changes with steroid hormone secretion. Endocrinology **60,** 641 (1957).

HAYNES, R. C., JR.: Activation of adrenal phosphorylase by adrenocorticotropic hormone. J. biol. Chem. **233,** 1220 (1958).

— BERTHET, L.: Studies on mechanism of action of adrenocorticotropic hormone. J. biol. Chem. **225,** 115 (1957).

— KORITZ, S. B., PERON, F. G.: Influence of adenosine 3',5'-monophosphate on corticoid production by rat adrenal glands. J. biol. Chem. **234,** 1421 (1959).

HAYSLETT, J. P., BOYD, J. E., EPSTEIN, F. H.: Aldosterone production in chronic renal failure. Proc. Soc. exp. Biol. (N.Y.) **130,** 912 (1969).

HELMER, O. M.: Studies on renin antibodies. Circulation **17,** 648 (1958).

— GRIFFITH, R. S.: The effect of the administration of estrogens on the renin-substrate (hypertensinogen) content of rat plasma. Endocrinology **51,** 421 (1952).

— JUDSON, W. E.: Influence of high renin substrate levels on renin-angiotensin system in pregnancy. Amer. J. Obstet. Gynec. **99,** 9 (1967).

— — Metabolic studies on hypertensive patients with suppressed plasma renin activity not due to hyperaldosteronism. Circulation **38,** 965 (1968).

HILF, R.: The mechanism of action of ACTH. New Eng. J. Med. **273,** 798 (1965).

HILTON, J. G., ALTER, S., CUSHMAN, P.: Aldosterone secretion: differences in direct effects of rubidium and potassium perfusion into dog adrenals. Nature (Lond.) **208,** 1099 (1965).

— KRUESI, O. R., NEDELJKOVIC, R. I., SCIAN, L. F.: Adrenocortical and medullary responses to adenosine 3',5'-monophosphate. Endocrinology **68,** 908 (1961).

— SCIAN, L. F., WESTERMANN, C. D., NAKANO, J., KRUESI, O. R.: Vasopressin stimulation of the isolated adrenal glands: nature and mechanism of hydrocortisone secretion. Endocrinology **67,** 298 (1960).

HOWARDS, S. S., DAVIS, J. O., JOHNSTON, C. I., WRIGHT, F. S.: Steroidogenic response in normal dogs receiving blood from dogs with caval constriction. Amer. J. Physiol. **214,** 990 (1968).

HUMPHREY, J. H., TOH, C. C.: Absorption of serotonin (5-hydroxytryptamine) and histamine by dog platelets. J. Physiol. (Lond.) **124,** 300 (1954).

IMURA, H., MATSUKURA, S., MATSUYAMA, H., SETSUDA, T., MIYAKE, T.: Adrenal steroidogenic effect of adenosine 3',5'-monophosphate and its derivatives *in vivo.* Endocrinology **76,** 933 (1965).

JENNY, M., MULLER, A. F., FABRE, J., MACH, R. S.: Hypokaliémie et alcalose par ingestion abusive d'extrait de réglisse (liquorice) et d'eau bicarbonatée. Pseudo-syndrome de Conn. Schweiz. med. Wschr. **91,** 869 (1961).

JESSIMAN, A. G., WATSON, D.D., MOORE, F. D.: Hypophysectomy in the treatment of breast cancer. New Engl. J. Med. **261,** 1199 (1959).

JOHNSON, B. B., LIEBERMAN, A. H., MULROW. P. J.: Aldosterone excretion in normal subjects depleted of sodium and potassium. J. clin. Invest. **36,** 757 (1957).

Johnston, C. I., Davis, J. O., Hartroft, P. M.: Renin-angiotensin system, adrenal steroids and sodium depletion in a primitive mammal, the American opossum. Endocrinology **81,** 633 (1967b).

— — Wright, F. S., Howards, S. S.: Effects of renin and ACTH on adrenal steroid secretion in the American bullfrog. Amer. J. Physiol. **213,** 393 (1967a).

Jones, K. M., Lloyd-Jones, R., Riondel, A. M., Tait, J. F., Tait, S. A. S., Bulbrook, R. D., Greenwood, F. C.: Aldosterone secretion and metabolism in normal men and women and in pregnancy. Acta endocr. (Kbh.) **30,** 321 (1959).

Jose, A., Kaplan, N. M.: Plasma renin activity in the diagnosis of primary aldosteronism. Failure to distinguish primary aldosteronism from essential hypertension. Arch. intern. Med. **123,** 141 (1969).

Jouan, P.: Epiphyse, 5-hydroxytryptamine et corticoidogenèse *in vitro.* Ann. Endocr. (Paris) **24,** 365 (1963).

— Propriétés adrénoglomérulotrophiques de la 5-hydroxytryptamine (sérotonine). Path. et Biol. **15,** 1145 (1967).

— Samperez, S.: Recherches sur la spécificité d'action de la 5-hydroxy-tryptamine vis-à-vis de la sécrétion *in vitro* de l'aldostérone. Ann. Endocr. (Paris) **25,** 70 (1964).

Kahnt, F. W., Neher, R.: On the specific inhibition of adrenal steroid biosynthesis. Experientia (Basel) **18,** 499 (1962).

— — Ueber die adrenale Steroid-Biosynthese *in vitro.* I. Umwandlung endogener und exogener Vorstufen im Nebennieren-Homogenat des Rindes. Helv. chim. Acta **48,** 1457 (1965).

— — Ueber die adrenale Steroid-Biosynthese *in vitro.* II. Bedeutung von Steroiden als Hemmstoffe. Helv. chim. Acta **49,** 123 (1966a).

— — Ueber die adrenale Steroid-Biosynthese *in vitro.* III. Selektive Hemmung der Nebennierenrinden-Funktion. Helv. chim. Acta **49,** 725 (1966b).

Kaplan, N. M.: The biosynthesis of adrenal steroids: effects of angiotensin II, adrenocorticotropin, and potassium. J. clin. Invest. **44,** 2029 (1965).

— The steroid content of adrenal adenomas and measurement of aldosterone production in patients with essential hypertension and primary aldosteronism. J. clin. Invest. **46,** 728 (1967).

— Bartter, F. C.: The effect of ACTH, renin, angiotensin II, and various precursors on biosynthesis of aldosterone by adrenal slices. J. clin. Invest. **41,** 715 (1962).

Karaboyas, G. C., Koritz, S. B.: Identity of the site of action of 3',5'-adenosine monophosphate and adrenocorticotropic hormone in corticosteroidogenesis in rat adrenal and beef adrenal cortex slices. Biochemistry **4,** 462 (1965).

Katz, F. H.: Primary aldosteronism with suppressed plasma renin activity due to bilateral nodular adrenocortical hyperplasia. Ann. intern. Med. **67,** 1035 (1967).

KAUFMANN, W., STEINER, B., DÜRR, F., NIETH, H., BEHN, C.: Aldo-steronstoffwechsel bei Nierenarterienstenose. Klin. Wschr. **45,** 966 (1967).

KINSON, G. A., SINGER, B.: The pineal gland and the adrenal response to sodium deficiency in the rat. Neuroendocrinology **2,** 283 (1967).

— — Sensitivity to angiotensin and adrenocorticotrophic hormone in the sodium deficient rat. Endocrinology **83,** 1108 (1968).

— — Influence of renal nerves on the secretion of aldosterone in the rat. Acta endocr. (Kbh.) **61,** 239 (1969).

— — GRANT, L.: Adrenocortical hormone secretion at various time intervals after pinealectomy in the rat. Gen. comp. Endocr. **10,** 447 (1968).

— WAHID, A. K., SINGER, B.: Effect of chronic pinealectomy on adreno-cortical hormone secretion rates in normal and hypertensive rats. Gen. comp. Endocr. **8,** 445 (1967).

KITTINGER, G. W.: Puromycin inhibition of *in vitro* cortical hormone production by the rat adrenal gland. Steroids **4,** 539 (1964).

KLIMAN, B., PETERSON, R. E.: Double isotope derivative assay of aldo-sterone in biological extracts. J. biol. Chem. **235,** 1639 (1960).

KOLPAKOV, M. G., KOLAEVA, S. G., KRASS, P. M., POLJAK, M. G., SOKO-LOVA, G. P., CHUDNOVSY, G. S., SHORIN, Y. P., STERENTAL, I. S.: Feedback mechanism in regulation of aldosterone and cortisol secretion. J. Steroid. Biochem. **1,** 111 (1970).

KORITZ, S. B.: On the regulation of pregnenolone synthesis. In: Functions of the Adrenal Cortex, Vol. 1. Ed.: K. W. McKERNS. Amsterdam: North Holland 1968, p. 27.

— PERON, F. G.: Studies on the mode of action of the adrenocorticotropic hormone. J. biol. Chem. **230,** 343 (1958).

KOWARSKI, A., FINKELSTEIN, J., LORAS, B., MIGEON, C. J.: The *in vitro* stability of the tritium label in 1,2-^{3}H-D-aldosterone when used for measurement of aldosterone secretion rate by the double isotope dilution technique. Steroids **3,** 95 (1964).

— RUSSELL, A., MIGEON, C. J.: Aldosterone secretion rate in the hyper-tensive form of congenital adrenal hyperplasia. J. clin. Endocr. **28,** 1445 (1968).

KRIEGER, D. T., KRIEGER, H. P.: Aldosterone excretion in pretectal disease. J. clin. Endocr. **24,** 1055 (1964).

KUMAR, D., FELTHAM, L. W., GORNALL, A. G.: Aldosterone excretion and tissue electrolytes in normal pregnancy and pre-eclampsia. Lancet **1959I,** 541.

LAIDLAW, J. C., COHEN, M., GORNALL, A. G.: Studies on the origin of aldosterone during human pregnancy. J. clin. Endocr. **18,** 222 (1958).

— RUSE, J. I., GORNALL, A. G.: The influence of estrogen and progesterone on aldosterone excretion. J. clin. Endocr. **22,** 161 (1962).

— YENDT, E. R., GORNALL, A. G.: Hypertension caused by renal artery occlusion simulating primary aldosteronism. Metabolism **9,** 612 (1960).

LAMBERG, B. A., PETTERSON, T., KARLSSON, R.: The effect of corticotropin and synthetic angiotensin on the glucose-6-phosphate and the succinic acid dehydrogenases in the adrenal cortex of the rat. Acta med. Scand. Suppl. **412,** 215 (1964).

LANDAU, R. L., BERGENSTAL, D. M., LUGIBIHL, K., KASCHT, M. E.: The metabolic effects of progesterone in man. J. clin. Endocr. **15,** 1194 (1955).

— LUGIBIHL, K.: Inhibition of the sodium-retaining influence of aldosterone by progesterone. J. clin. Endocr. **18,** 1237 (1958).

— — DIMICK, D. F.: Metabolic effects in man of steroids with progestational activity. Ann. N.Y. Acad. Sci. **71,** 588 (1958).

LARAGH, J. H., ANGERS, M., KELLY, W. G., LIEBERMAN, S.: Hypotensive agents and pressor substances. The effect of epinephrine, norepinephrine, angiotensin II and others on the secretory rate of aldosterone in man. J. Amer. med. Ass. **174,** 234 (1960a).

— KELLY, W. G.: Aldosterone: its biochemistry and physiology. In: Advances in Metabolic Disorders, Vol. 1. Eds.: R. LEVINE and R.LUFT. New York: Academic Press 1964, p. 217.

— SEALEY, J. E., LEDINGHAM, J. G. G., NEWTON, M. A.: Oral contraceptives. Renin, aldosterone, and high blood pressure. J. Amer. med. Ass. **201,** 918 (1967).

— — SOMMERS, S. C.: Patterns of adrenal secretion and urinary excretion of aldosterone and plasma renin activity in normal and hypertensive subjects. Circulation Res. Suppl. 1 to Vols. **18** and **19,** 158 (1966).

— STOERK, H. C.: A study of the mechanism of secretion of the sodium-retaining hormone (aldosterone). J. clin. Invest. **36,** 383 (1957).

— ULICK, S., JANUSZEWICZ, V., DEMING, Q. B., KELLY, W. G., LIEBERMAN, S.: Aldosterone secretion and primary and malignant hypertension. J. clin. Invest. **39,** 1091 (1960b).

LAYNE, D. S., MEYER, C. J., VAISHWANAR, P. S., PINCUS, G.: The secretion and metabolism of cortisol and aldosterone in normal and in steroid-treated women. J. clin. Endocr. **22,** 107 (1962).

LEE, T. C., BIGLIERI, E. G., VAN BRUNT, E. E., GANONG, W. F.: Inhibition of aldosterone secretion by passive transfer of antirenin antibodies to dogs on a low sodium diet. Proc. Soc. exp. Biol. (N.Y.) **119,** 315 (1965).

— DE WIED, D.: Somatotropin as the non-ACTH factor of anterior pituitary origin for the maintenance of enhanced aldosterone secretory responsiveness of dietary sodium restriction in chronically hypophysectomized rats. Life Sci. **7,** 35 (1968).

— VAN DER WAL, B., DE WIED, D.: Influence of the anterior pituitary on the aldosterone secretory response to dietary sodium restriction in the rat. J. Endocr. **42,** 465 (1968).

LIDDLE, G. W.: Effects of anti-inflammatory steroids on electrolyte metabolism. Ann. N.Y. Acad. Sci. **82,** 797 (1959).

— DUNCAN, L. E., JR., BARTTER, F. C.: Dual mechanism regulating adrenocortical function in man. Amer. J. Med. **21,** 380 (1956).

LIEBERMAN, A. H., LUETSCHER, J. A., JR.: Some effects of abnormalities of pituitary, adrenal or thyroid function on excretion of aldosterone and the response to corticotropin or sodium deprivation. J. clin. Endocr. **20,** 1004 (1960).

LOMMER, D.: Hemmung der Corticosteroid-11β-hydroxylierung durch einen Extrakt aus corpus pineale. Experientia (Basel) **22,** 122 (1966).

— DÜSTERDIECK, G., JAHNECKE, J., VECSEI, P., WOLFF, H. P.: Sekretion, Plasmakonzentration, Verteilung, Stoffwechsel und Ausscheidung von Aldosteron bei Gesunden und Kranken. Klin. Wschr. **46,** 741 (1968).

— WOLFF, H. P.: ACTH-Stimulierung der Corticosteroidbiosynthese durch Freisetzung „gebundener" 21-Desoxypregnene. Experientia (Basel) **22,** 654 (1966a).

— — Stimulation of the *in vitro* biosynthesis of corticosteroids by angiotensin II. Experientia (Basel) **22,** 699 (1966b).

LOUIS, L. H., CONN, J. W.: Primary aldosteronism: content of adrenocortical steroids in adrenal tissue. Recent Progr. Hormone Res. **17,** 415 (1961).

LUCIS, O. J., DYRENFURTH, I., VENNING, E. H.: Effect of various preparations of pituitary and diencephalon on the *in vitro* secretion of aldosterone and corticosterone by the rat adrenal gland. Canad. J. Biochem. **39,** 901 (1961).

— VENNING, E. H.: *In vitro* and *in vivo* effect of growth hormone on aldosterone secretion. Canad. J. Biochem. **38,** 1069 (1960).

LUCIS, R., CARBALLEIRA, A., VENNING, E. H.: Biotransformation of progesterone-4-14C and 11-deoxycorticosterone-4-14C by adrenal glands *in vitro*. Steroids **6,** 737 (1965).

LUETSCHER, J. A., JR., AXELRAD, B. J.: Increased aldosterone output during sodium deprivation in normal men. Proc. Soc. exp. Biol. (N.Y.) **87,** 650 (1954).

MACCHI, I. A.: *In vitro* action of mammalian adrenocorticotropin and 5-hydroxytryptamine on adrenocortical secretion in the turtle, snake, and bullfrog. Amer. Zool. (abstract) **3,** 548 (1963).

MARIEB, N. J., MULROW, P. J.: Role of the renin-angiotensin system in the regulation of aldosterone secretion in the rat. Endocrinology **76,** 657 (1965).

MARUSIC, E. T., MULROW, P. J.: *In vitro* conversion of corticosterone-4-14C to 18-hydroxycorticosterone by zona fasciculata-reticularis of beef adrenal. Endocrinology **80,** 214 (1967a).

— — Stimulation of aldosterone biosynthesis in adrenal mitochondria by sodium depletion. J. clin. Invest. **46,** 2101 (1967b).

MARX, A. J., DEANE, H. W.: Histophysiologic changes in the kidney and adrenal cortex in rats on a low-sodium diet. Endocrinology **73,** 317 (1963).

— — MOWLES, T. F., SHEPPARD, H.: Chronic administration of angiotensin in rats; changes in blood pressure, renal and adrenal histophysiology and aldosterone production. Endocrinology **73,** 329 (1963).

Masson, G. M. C., Travis, R. H.: Effects of renin on aldosterone secretion in the rat. Canad. J. Physiol. Pharmacol. **46,** 11 (1968).

McCaa, C. S., Richardson, T. Q., McCaa, R. E., Sulya, L. L., Guyton, A. C.: Aldosterone secretion by dogs during the development phase of Goldblatt hypertension. J. Endocr. **33,** 97 (1965).

McKenzie, J. K., Montgomerie, J. Z.: Renin-like activity in the plasma of anephric men. Nature (Lond.) **223,** 1156 (1969).

McKerns, K. W.: Mechanisms of ACTH regulation of the adrenal cortex. In: Functions of the Adrenal Cortex, Vol. 1. Ed.: K. W. McKerns. Amsterdam: North Holland 1968, p. 479.

Melby, J. C., Wilson, T. E., Dale, S. L., Hickler, R. V.: Extrarenal regulation of aldosterone secretion in primary aldosteronism. J. clin. Invest. (abstract) **44,** 1074 (1965).

Meyer, C. J., Layne, D. S., Tait, J. F., Pincus, G.: The binding of aldosterone to plasma proteins in normal, pregnant, and steroid-treated women. J. clin. Invest. **40,** 1663 (1961).

Miura, K., Yoshinaga, K., Goto, K., Katsushima, I., Maebashi, M., Demura, H., Iino, M., Demura, R., Torikai, T.: A case of glucocorticoid-responsive hyperaldosteronism. J. clin. Endocr. **28,** 1807 (1968).

Müller, J.: Aldosterone stimulation *in vitro*. I. Evaluation of assay procedure and determination of aldosterone-stimulating activity in a human urine extract. Acta endocr. (Kbh.) **48,** 283 (1965a).

— Stimulation of aldosterone production *in vitro* by ammonium ions. Nature (Lond.) **206,** 92 (1965b),

— Aldosterone stimulation *in vitro*. II. Stimulation of aldosterone production by monovalent cations. Acta endocr. (Kbh.) **50,** 301 (1965c).

— Aldosterone stimulation *in vitro*. III. Site of action of different aldosterone-stimulating substances on steroid biosynthesis. Acta endocr. (Kbh.) **52,** 515 (1966).

— Alterations of aldosterone biosynthesis by rat adrenal tissue due to increased intake of sodium and potassium. Acta endocr. (Kbh.) **58,** 27 (1968).

— Decreased aldosterone production by rat adrenal tissue *in vitro* due to treatment with 9α-fluorocortisol, dexamethasone and adrenocorticotrophin *in vivo*. Acta endocr. (Kbh.) **63,** 1 (1970).

— Steroidogenic effect of stimulators of aldosterone biosynthesis upon separate zones of the rat adrenal cortex. Influence of sodium and potassium deficiency. Europ. J. clin. Invest. (in press, 1971).

— Gross, F.: Effects of experimental renal hypertension on aldosterone biosynthesis by rat adrenal tissue. Acta endocr. (Kbh.) **60,** 669 (1969).

— Huber, R.: Effects of sodium deficiency, potassium deficiency, and uremia upon the steroidogenic response of rat adrenal tissue to serotonin, potassium ions, and adrenocorticotropin. Endocrinology **85,** 43 (1969).

— Weick, W. J.: Aldosterone stimulation *in vitro*. IV. Corticosteroid-stimulating activity in rat serum. Acta endocr. (Kbh.) **54,** 63 (1967).

MÜLLER, J., ZIEGLER, W. H.: Stimulation of aldosterone biosynthesis *in vitro* by serotonin. Acta endocr. (Kbh.) **59,** 23 (1968).

MULLER, A. F.: Regulation der Aldosteronsekretion. Verh. Dtsch. Ges. inn. Med. **68,** 599 (1963).

— MANNING, E. L., MORET, P., MEGEVAND, R.: Renal blood supply and aldosterone secretion. In: Aldosterone. Eds.: E. E. BAULIEU and P. ROBEL. Oxford: Blackwell 1964, p. 187.

— — RIONDEL, A. M.: L'excrétion de l'aldostérone chez le sujet normal pendant la déplétion et la réplétion en potassium. Helv. med. Acta **25,** 547 (1958).

— RIONDEL, A. M., MANNING, E. L.: Effect of corticotrophin on secretion of aldosterone. Lancet **1956 II,** 1021.

MULROW, P. J.: Neural and other mechanisms regulating aldosterone secretion. In: Neuroendocrinology, Vol. 1. Eds.: L. MARTINI and W. F. GANONG. New York: Academic Press 1966, p. 407.

— COHN, G. L.: Conversion of corticosterone-4-^{14}C to aldosterone by human adrenal slices. Proc. Soc. exp. Biol. (N.Y.) **101,** 731 (1959).

— GANONG, W. F.: The effect of hemorrhage upon aldosterone secretion in normal and hypophysectomized dogs. J. clin. Invest. **40,** 579 (1961).

— — BORYCZKA, A.: Further evidence for a role of the renin-angiotensin system in control of aldosterone secretion. Proc. Soc. exp. Biol. (N.Y.) **112,** 7 (1963).

— — CERA, G., KULJIAN, A.: The nature of the aldosterone stimulating factor in dog kidneys. J. clin. Invest. **41,** 505 (1962).

NEHER, R.: Aldosterone and other adrenocortical hormones in human adrenals and adrenal tumours. In: An International Symposium on Aldosterone. Eds.: A. F. MULLER and C. M. O'CONNOR. Boston: Little, Brown & Co. 1958, p. 11.

NEW, M. I., MILLER, B., PETERSON, R. E.: Aldosterone excretion in normal children and in children with adrenal hyperplasia. J. clin. Invest. **45,** 412 (1966).

— PETERSON, R. E.: A new form of congenital adrenal hyperplasia. J. clin. Endocr. **27,** 300 (1967).

NEWBORG, B., KEMPNER, W.: Analysis of 177 cases of hypertensive vascular disease with papilledema. One hundred twenty-six patients treated with rice diet. Amer. J. Med. **19,** 33 (1955).

NEWTON, M. A., LARAGH, J. H.: Effect of corticotropin on aldosterone excretion and plasma renin in normal subjects, in essential hypertension and in primary aldosteronism. J. clin. Endocr. **28,** 1006 (1968 a).

— — Effects of glucocorticoid administration on aldosterone excretion and plasma renin in normal subjects, in essential hypertension and in primary aldosteronism. J. clin. Endocr. **28,** 1014 (1968 b).

NEY, R. L.: Effects of dibutyryl cyclic AMP on adrenal growth and steroidogenic capacity. Endocrinology **84,** 168 (1969).

NICOLIS, G. L., ULICK, S.: Role of 18-hydroxylation in the biosynthesis of aldosterone. Endocrinology **76,** 514 (1965).

OMURA, T., SATO, R., COOPER, D. Y., ROSENTHAL, O., ESTABROOK, R. W.: Function of cytochrome P-450 of microsomes. Fed. Proc. **24,** 1181 (1965).

PALKOVITS, M., FÖLDVARI, P. I.: Effect of the subcommissural organ and the pineal body on the adrenal cortex. Endocrinology **72,** 28 (1963).

— MONOS, E., FACHET, J.: The effect of subcommissural-organ lesions on aldosterone production in the rat. Acta endocr. (Kbh.) **48,** 169 (1965).

PALMORE, W. P., ANDERSON, R., MULROW, P. J.: Role of the pituitary in controlling aldosterone production in sodium-depleted rats. Endocrinology **86,** 728 (1970).

— MARIEB, N. J., MULROW, P. J.: Stimulation of aldosterone secretion by sodium depletion in nephrectomized rats. Endocrinology **84,** 1342 (1969).

— MULROW, P. J.: Control of aldosterone secretion by the pituitary gland. Science **158,** 1482 (1967).

PASQUALINI, J. R.: Conversion of tritiated 18-hydroxycorticosterone to aldosterone by slices of human cortico-adrenal gland and adrenal tumour. Nature (Lond.) **201,** 501 (1964).

PEART, W. S.: The renin-angiotensin system. Pharmacol. Rev. **17,** 143 (1965).

PERON, F. G., KORITZ, S. B.: On the exogenous requirements for the action of ACTH *in vitro* on rat adrenal glands. J. biol. Chem. **233,** 256 (1958).

PESCHEL, E., RACE, G. J.: Studies on the adrenal zona glomerulosa of hypertensive patients and rats. Amer. J. Med. **17,** 355 (1954).

PETERSON, R. E.: Plasma corticosterone and hydrocortisone levels in man. J. clin. Endocr. **17,** 1150 (1957).

— The miscible pool and turnover rate of adrenocortical steroids in man. Recent Progr. Hormone Res. **15,** 231 (1959).

— NOKES, G., CHEN, P. S., BLACK, R. L.: Estrogens and adrenocortical function in man. J. clin. Endocr. **20,** 495 (1960).

PREEDY, J. R. K., AITKEN, E. H.: The effect of estrogen on water and electrolyte metabolism. I. The normal. J. clin. Invest. **35,** 423 (1956a).

— — The effect of estrogen on water and electrolyte metabolism. II. Hepatic disease. J. clin. Invest. **35,** 430 (1956b).

PSYCHOYOS, S., TALLAN, H. H., GREENGARD, P.: Aldosterone synthesis by adrenal mitochondria. J. biol. Chem. **241,** 2949 (1966).

RAMAN, P. B., SHARMA, D. C., DORFMAN, R. I.: Studies on aldosterone biosynthesis *in vitro*. Biochemistry **5,** 1795 (1966).

RAND, M., REID, G.: Source of serotonin in serum. Nature (Lond.) **168,** 385 (1951).

RAPP, J. P.: Adrenal steroidogenesis and serum renin in rats bred for juxtaglomerular granularity. Amer. J. Physiol. **216,** 860 (1969a).

— Deoxycorticosterone production in adrenal regeneration hypertension: *in vitro vs. in vivo* comparison. Endocrinology **84,** 1409 (1969b).

RAPPAPORT, R., DRAY, F., LEGRAND, J. C., ROYER, P.: Hypoaldostéronisme congénital familial par défaut de la 18-OH-déhydrogénase. Pediat. Res. **2,** 456 (1968).

REMBIESA, R., YOUNG, P. C. M., SAFFRAN, M.: Effect of 17α-methyl-testosterone on steroid formation by rat adrenal tissue. Canad. J. Biochem. **46,** 433 (1968).

RICHTER, C. L., VEILLEUX, R., BOIS, P.: Effect of magnesium deficiency on corticosterone in rats. Endocrinology **82,** 954 (1968).

ROBB, C. A., DAVIS, J. O., JOHNSTON, C. I., HARTROFT, P. M.: Effects of deoxycorticosterone on plasma renin activity in conscious dogs. Amer. J. Physiol. **216,** 884 (1969).

ROSENFELD, G., ROSEMBERG, E., UNGAR, F., DORFMAN, R. I.: Regulation of the secretion of aldosterone-like material. Endocrinology **58,** 255 (1956).

ROSENKRANTZ, H.: A direct influence of 5-hydroxytryptamine on the adrenal cortex. Endocrinology **64,** 355 (1959).

— LAFERTE, R. O.: Further observations on the relationship between serotonin and the adrenal. Endocrinology **66,** 832 (1960).

ROSNAGLE, R. S., FARRELL, G. L.: Alterations in electrolyte intake and adrenal steroid secretion. Amer. J. Physiol. **187,** 7 (1957).

ROSS, E. J.: Conn's syndrome due to adrenal hyperplasia with hypertrophy of zona glomerulosa, relieved by unilateral adrenalectomy. Amer. J. Med. **39,** 994 (1965).

SAFFRAN, M., SCHALLY, A. V.: *In vitro* bioassay of corticotropin: modi-fication and statistical treatment. Endocrinology **56,** 523 (1955).

SALASSA, R. M., MATTOX, V. R., ROSEVEAR, J. W.: Inhibition of the „mineralocorticoid" activity of licorice by spironolactone. J. clin. Endocr. **22,** 1156 (1962).

SALTI, I. S., RUSE, J. L., STIEFEL, M., DELARUE, N. C., LAIDLAW, J. C.: Non-tumorous „primary" aldosteronism: II. Type not relieved by glucocorticoid. Canad. med. Ass. J. **101,** 11 (1969).

SAMBHI, M. P., LEVITAN, B. A., BECK, J. C., VENNING, E. H.: The rate of aldosterone secretion in hypertensive patients with demonstrable renal artery stenosis. Metabolism **12,** 498 (1963).

SCHAECHTELIN, G., REGOLI, D., GROSS, F.: Quantitative assay and dis-appearance rate of circulating renin. Amer. J. Physiol. **206,** 1361 (1964).

SCHATZMANN, H. J.: Herzglykoside als Hemmstoffe für den aktiven Kalium- und Natriumtransport durch die Erythrocytenmembran. Helv. physiol. pharmacol. Acta **11,** 346 (1953).

SCHLATMANN, R. J. A. F. M., JANSEN, A. P., PRENEN, H., VAN DER KORST, J. K., MAJOOR, C. L. H.: The natriuretic and aldosterone-suppressive action of heparin and some related polysulfated polysaccharides. J. clin. Endocr. **24,** 35 (1964).

SCHRIEFERS, H., BAYER, J. M., PITTEL, M.: Vergleichende Untersuchun-gen zur Biogenese von Steroidhormonen bei Durchströmung über-lebender Nebennieren einer Patientin mit Cushing- und einer Patientin mit Conn-Syndrom. Acta endocr. (Kbh.) **43,** 419 (1963).

SCORNIK, O. A., PALADINI, A. C.: Significance of blood angiotensin levels in different experimental conditions. Canad. med. Ass. J. **90,** 269 (1964).

SHARMA, D. C.: Studies on aldosterone biosynthesis *in vitro*. Acta endocr. (Kbh.) **63,** 299 (1970).

— NERENBERG, C. A., DORFMAN, R. I.: Studies on aldosterone biosynthesis *in vitro*. II. Biochemistry **6,** 3472 (1967).

SHARP, G. W. G., LEAF, A.: Mechanism of action of aldosterone. Physiol. Rev. **46,** 593 (1966).

SHEPPARD, H., MOWLES, T. F., CHART, J. J., RENZI, A. A., HOWIE, N.: Steroid biosynthesis by rat adrenal: during development of adrenal regeneration and desoxycorticosterone acetate-induced hypertension. Endocrinology **74,** 762 (1964).

— SWENSON, R., MOWLES, T. F.: Steroid biosynthesis by rat adrenal: functional zonation. Endocrinology **73,** 819 (1963).

SINGER, B., LOSITO, C., SALMON, S.: Aldosterone and corticosterone secretion rates in rats with experimental renal hypertension. Acta endocr. (Kbh.) **44,** 505 (1963a).

— — — Effect of progesterone on adrenocortical hormone secretion in normal and hypophysectomized rats. J. Endocr. **28,** 65 (1963b).

— STACK-DUNNE, M. P.: The secretion of aldosterone and corticosterone by the rat adrenal. J. Endocr. **12,** 130 (1955).

SLATER, J. D. H., BARBOUR, B. H., HENDERSON, H. H., CASPER, A. G. T., BARTTER, F. C.: Physiological influence of the kidney on the secretion of aldosterone, corticosterone and cortisol by the adrenal cortex. Clin. Sci. **28,** 219 (1965).

— TUFFLEY, R. E., WILLIAMS, E. S., BERESFORD, C. H., SÖNKENS, P. H., EDWARDS, R. H. T., EKINS, R. P., MCLAUGHLIN, M.: Control of aldosterone secretion during acclimatization to hypoxia in man. Clin. Sci. **37,** 327 (1969).

SLATON, P. E., JR., BIGLIERI, E. G.: Hypertension and hyperaldosteronism of renal and adrenal origin. Amer. J. Med. **38,** 324 (1964).

— SCHAMBELAN, M., BIGLIERI, E. G.: Stimulation and suppression of aldosterone secretion in patients with an aldosterone-producing adenoma. J. clin. Endocr. **29,** 239 (1969).

SOLOMON, S. S., SWERSIE, S. P., PAULSEN, C. A., BIGLIERI, E. G.: Feminizing adrenocortical carcinoma with hypertension. J. clin. Endocr. **28,** 608 (1968).

SPÄT, A., SOLYOM, J., STURCZ, J., MESZAROS, I., LUDWIG, E.: Effect of angiotensin superfusion on the rate of aldosterone production by incubated rat adrenals. Acta physiol. Acad. Sci. Hung. **35,** 149 (1969).

— STURCZ, J., BANKI, L., SZIGETI, R.: Effect on adrenal steroid secretion of changes in renal haemodynamics in the rat. Acta physiol. Acad. Sci. Hung. **30,** 209 (1966b).

— — SZIGETI, R.: New observations on the function of the angiotensin-aldosterone system. Acta physiol. Acad. Sci. Hung. **27,** 199 (1965).

Spät, A., Solyom, J., Szigeti, R., Banki, L.: Study on the role of the kidney in the regulation of aldosterone secretion in the rat. Endokrinologie 50, 276 (1966a).

Spark, R. F., Gordon, S. J., Dale, S. L., Melby, J. C.: Aldosterone production after suppression of corticotropic secretory activity. Arch. intern. Med. 122, 394 (1968).

Stachenko, J., Giroud, C. J. P.: Functional zonation of the adrenal cortex: site of ACTH action. Endocrinology 64, 743 (1959).

— — Further observations on the functional zonation of the adrenal cortex. Canad. J. Biochem. 42, 1777 (1964).

–— Laplante, C., Giroud, C. J. P.: Double isotope derivative assay of aldosterone, corticosterone, and cortisol. Canad. J. Biochem. 42, 1275 (1964).

Stark, G.: Zum Aldosteron- und Elektrolyt-Stoffwechsel in der Schwangerschaft. Arch. Gynäk. 203, 192 (1966).

— Kossmann, H.: Die Ausscheidung von Aldosteron, Natrium und Kalium nach fortlaufender hoher Progesterongabe. Acta endocr. (Kbh.) 42, 537 (1963).

Steinacker, H. G., Vecsei, P., Lommer, D., Wolff, H. P.: In vitro corticosteroid biosynthesis in adrenals of rats uraemic after bilateral nephrectomy. Acta endocr. (Kbh.) 58, 630 (1968).

Stone, D., Hechter, O.: Studies on ACTH action in perfused bovine adrenals: The site of action of ACTH in corticosteroidogenesis. Arch. Biochem. 51, 457 (1954).

Sturcz, J., Spät, A., Szöcs, G., Köver, G., Farkas, A.: Effect of potassium on adrenal aldosterone production in rats. Acta endocr. (Kbh.) 55, 193 (1967).

Sutherland, D. J. A., Ruse, J. L., Laidlaw, J. C.: Hypertension, increased aldosterone secretion and low plasma renin activity relieved by dexamethasone. Canad. med. Ass. J. 95, 1109 (1966).

Sutherland, E. W., Øye, I., Butcher, R. W.: The action of epinephrine and the role of the adenyl cyclase system in hormone action. Recent Progr. Hormone Res. 21, 623 (1965).

Swann, H. G.: Pituitary-adrenocortical relationship. Physiol. Rev. 20, 493 (1940).

Tait, J. F., Little, B., Tait, S. A. S., Flood, C.: The metabolic clearance rate of aldosterone in pregnant and non pregnant subjects estimated by both single-injection and constant-infusion methods. J. clin. Invest. 41, 2093 (1962).

Tait, S. A. S., Schulster, D., Okamoto, M., Flood, C., Tait, J. F.: Production of steroids by in vitro superfusion of endocrine tissue. II. Steroid output from bisected whole, capsular and decapsulated adrenals of normal intact, hypophysectomized and hypophysectomized-nephrectomized rats as a function of time of superfusion. Endocrinology 86, 360 (1970).

Tallan, H. H., Psychoyos, S., Greengard, P.: Aldosterone synthesis by adrenal mitochondria. II. The effect of citric acid cycle intermediates;

identification of the soluble stimulatory factor as fumarase. J. biol. Chem. **242,** 1912 (1967).

TAYLOR, A. N., FARRELL, G.: Effect of brain stem lesions on aldosterone and cortisol secretion. Endocrinology **70,** 556 (1962).

TELEGDY, G., LISSAC, K.: The effect of progesterone on adrenal corticosterone and aldosterone secretion in the female rat. Acta physiol. Acad. Sci. Hung. **26,** 313 (1965).

THORN, G. W., ROSS, E. J., CRABBE, J., VAN'T HOFF, W.: Studies on aldosterone secretion in man. Brit. med. J. **1957**II, 955.

TOUITOU, Y., MALMEJAC, A., AUPETIT, B., BIANCANI, J., LEGRAND, J.-C.: Taux de sécrétion de 18-hydroxycorticostérone. Application de la méthode d'Ulick modifiée à différents cas cliniques. Ann. Biol. clin. **28,** 49 (1970).

TRAIKOV, H., DE NICOLA, A. F., BIRMINGHAM, M. K.: The effects of 2-methyl-1,2-bis(3'pyridyl)-1-propanol (reduced Metopirone) on the formation of 18-hydroxylated steroids by quartered rat adrenals. Steroids **13,** 457 (1969).

TUCCI, J. R., ESPINER, E. A., JAGGER, P. I., LAULER, D. P.: The effect of Metyrapone on aldosterone secretion in man. Acta endocr. (Kbh.) **56,** 376 (1967b).

— — — PAUK, G. L., LAULER, D. P.: ACTH stimulation of aldosterone secretion in normal subjects and in patients with chronic adrenocortical insufficiency. J. clin. Endocr. **27,** 568 (1967a).

ULICK, S., FEINHOLTZ, E.: Metabolism and rate of secretion of aldosterone in the bullfrog. J. clin. Invest. **47,** 2523 (1968).

— GAUTIER, E., VETTER, K. K., MARKELLO, J. R., YAFFE, S., LOWE, C. U.: An aldosterone biosynthetic defect in a salt-losing disorder. J. clin. Endocr. **24,** 669 (1964b).

— NICOLIS, G. L., VETTER, K. K.: Relationship of 18-hydroxycorticosterone to aldosterone. In: Aldosterone. Eds.: E. E. BAULIEU and P. ROBEL. Oxford: Blackwell 1964a, p. 3.

— VETTER, K. K.: Simultaneous measurement of secretory rates of aldosterone and 18-hydroxycorticosterone. J. clin. Endocr. **25,** 1015 (1965).

URQUHART, J., DAVIS, J. O., HIGGINS, J. T., JR.: Effects of prolonged infusion of angiotensin II in normal dogs. Amer. J. Physiol. **205,** 1241 (1963).

— — — Simulation of spontaneous secondary hyperaldosteronism by intravenous infusion of angiotensin II in dogs with an arteriovenous fistula. J. clin. Invest. **43,** 1355 (1964).

VECSEI, P., LOMMER, D., STEINACKER, H. G., VECSEI-GÖRGENYI, A., WOLFF, H. P.: *In vitro* Corticosteroidbiosynthese in proliferierenden Rattennebennieren. Acta endocr. (Kbh.) **53,** 24 (1966a).

— — — WOLFF, H. P.: Changes in 18-hydroxycorticosterone and aldosterone synthesis in rat adrenals *in vitro* after renal hypertension, nephrectomy, and variation of sodium intake *in vivo*. Eur. J. Steroids **1,** 91 (1966b).

VECSEI, P., LOMMER, D., WOLFF, H. P.: The intermediate role of 18-hydroxy-corticosteroids in aldosterone biosynthesis. Experientia (Basel) **24,** 1199 (1968).

VENNING, E. H., DYRENFURTH, I., BECK, J. C.: Effect of corticotropin and prednisone on the excretion of aldosterone in man. J. clin. Endocr. **16,** 1541 (1956).

— — DOSSETOR, J. B., BECK, J. C.: Influence of alterations in sodium intake on urinary aldosterone response to corticotropin in normal individuals and patients with essential hypertension. Metabolism **11,** 254 (1962).

— — LOWENSTEIN, L., BECK, J.: Metabolic studies in pregnancy and the puerperium. J. clin. Endocr. **19,** 403 (1959).

— LUCIS, O. J.: Effect of growth hormone on the biosynthesis of aldosterone in the rat. Endocrinology **70,** 486 (1962).

VERDESCA, A. S., WESTERMANN, C. D., CRAMPTON, R. S., BLACK, W. C., NEDELJKOVIC, R. I., HILTON, J. G.: Direct adrenocortical stimulatory effect of serotonin. Amer. J. Physiol. **201,** 1065 (1961).

VEYRAT, R., BRUNNER, H. R., GRANDCHAMP, A., MULLER, A. F.: Inhibition of renin by potassium in man. (abstract) Acta endocr. (Kbh.) Suppl. **119,** 86 (1967).

— MANNING, E. L., FABRE, J., MULLER, A. F.: Mesure de la sécrétion de l'aldostérone sous administration d'un adrénostatique semi-synthétique, l'héparinoïde Ro 1—8307. Rev. Franç. Étud. clin. biol. **8,** 667 (1963).

— MULLER, A. F., MACH, R. S.: Les hyperaldostéronismes secondaires. Schweiz. med. Wschr. **98,** 65 (1968).

VINSON, G. P., WHITEHOUSE, B. J.: The relationship between steroid production from endogenous precursors and from added radioactive precursors by rat adrenal tissue *in vitro*. The effect of corticotrophin. Acta endocr. (Kbh.) **61,** 695 (1969).

VISSER, H. K. A., COST, W. S.: A new hereditary defect in the biosynthesis of aldosterone: urinary C_{21}-corticoid pattern in three related patients with a salt-losing syndrome suggesting a 18-oxidation defect. Acta endocr. (Kbh.) **47,** 589 (1964).

WAKERLIN, G. E.: Antibodies to renin as proof of the pathogenesis of sustained renal hypertension. Circulation **17,** 653 (1958).

VAN DER WAL, B., DE WIED, D.: Effects of various peptides on aldosterone production of sodium deficient rats. Acta endocr. (Kbh.) **59,** 186 (1968).

WATANABE, M., MEEKER, C. I., GRAY, M. J., SIMS, E. A. H., SOLOMON, S.: Secretion rate of aldosterone in normal pregnancy. J. clin. Invest. **42,** 1619 (1963).

— — — — — Aldosterone secretion rates in abnormal pregnancy. J. clin. Endocr. **25,** 1665 (1965).

WEIDMANN, P., SIEGENTHALER, W.: Das Renin-Angiotensin-Aldosteron-System bei hypertensiven Zuständen. Dtsch. med. Wschr. **92,** 1953 (1967).

WEIDMANN, P., SIEGENTHALER, W., ZIEGLER, W., SULSER, H., ENDRES, P., WERNING, C.: Hypertension associated with tumors adjacent to renal arteries. Amer. J. Med. **47,** 528 (1969).

WELLEN, J. J., BENRAAD, T. J.: Effect of ouabain on corticosterone biosynthesis and on potassium and sodium concentration in calf adrenal tissue *in vitro*. Biochim. biophys. Acta (Amst.) **183,** 110 (1969a).

— — Effect of ouabain on corticosterone biosynthesis and on intracellular potassium and sodium concentration in adrenal tissue *in vitro*. Acta endocr. (Kbh.) (abstract) Suppl. **138,** 50 (1969b).

WERNING, C., BAUMANN, K., WEIDMANN, P., GYSLING, E., SIEGENTHALER, W.: Die Plasmareninaktivität bei dekompensiertem und nicht-dekompensiertem Diabetes insipidus. Zugleich ein Beitrag zur Regulation der Reninsekretion. Schweiz. med. Wschr. **99,** 661 (1969).

WETTSTEIN, A.: Biosynthèse des hormones stéroides. Experientia (Basel) **17,** 329 (1961).

WHITE, F. N., GOLD, E. M., VAUGHN, D. L.: Renin-aldosterone system in endotoxin shock in the dog. Amer. J. Physiol. **212,** 1195 (1967).

WILLIAMS, G. H., ROSE, L. I., JAGGER, P. I., LAULER, D. P.: Abnormal aldosterone response to acute stimulation in hypopituitarism. Clin. Res. (abstract) **16,** 526 (1968a).

— — — — Normal aldosterone secretory response in patients with long-term steroid suppression. Clin. Res. (abstract) **16,** 527 (1968b).

WILSON, I. D., GOETZ, F. C.: Selective hypoaldosteronism after prolonged heparin administration. A case report, with postmortem findings. Amer. J. Med. **36,** 635 (1964).

WOLFE, L. K., GORDON, R. D., ISLAND, D. P., LIDDLE, G. W.: An analysis of factors determining the circadian pattern of aldosterone excretion. J. clin. Endocr. **26,** 1261 (1966).

WOMERSLEY, R. A., DARRAGH, J. H.: Potassium and sodium restriction in the normal human. J. clin. Invest. **34,** 456 (1955).

WRONG, O.: Communication to the 51st annual general meeting of the Association of Physicians of Great Britain and Ireland, 1957. Quart. J. Med. **26,** 586 (1957).

— Hyperaldosteronism secondary to renal ischaemia. In: Aldosterone. Eds.: E. E. BAULIEU and P. ROBEL. Oxford: Blackwell 1964, p. 377.

Acknowledgements

All my own research cited in this review has been carried out in the Steroid Laboratory, Metabolic Unit, Department of Medicine, University of Zurich at the Kantonsspital Zurich and was supported by Grants 2579, 3853 and 4742.3 from the Schweizerische National-fonds. Some of these studies were performed in collaboration with Dr. W. Joe Weick, Dr. Walter H. Ziegler, Dr. Rolf Huber and Prof. Franz Gross. I am most grateful for the excellent technical assistance by Mrs. Daniella Vogelsang, Miss Catharina Modéer, Mrs. Gisela Möhren, Miss Lotti Berchtold and Mrs. Elsbeth Läuffer. I wish to thank Dr. J. P. Coghlan, Prof. E. R. Froesch and Prof. A. Labhart for their critical advice and continued support and encouragement. I am particularly grateful to Dr. Ralph E. Peterson, who initiated my studies on the regulation of aldosterone biosynthesis and who was my major instructor in steroid biochemistry. I further acknowledge the careful and efficient secretarial work by Miss M. Schittenhelm and Mrs. C. Bühler.

Typesetting, printing and binding: Konrad Triltsch, Graphischer Betrieb, 87 Würzburg, Germany

Monographs on Endocrinology